Applications of Photogrammetry for Environmental Research

Applications of Photogrammetry for Environmental Research

Special Issue Editors

Francesco Mancini
Riccardo Salvini

MDPI • Basel • Beijing • Wuhan • Barcelona • Belgrade

Special Issue Editors

Francesco Mancini
University of Modena and Reggio Emilia
Italy

Riccardo Salvini
University of Siena
Italy

Editorial Office
MDPI
St. Alban-Anlage 66
4052 Basel, Switzerland

This is a reprint of articles from the Special Issue published online in the open access journal *ISPRS International Journal of Geo-Information* (ISSN 2220-9964) in 2019 (available at: https://www.mdpi.com/journal/ijgi/special_issues/photogrammetry_environment).

For citation purposes, cite each article independently as indicated on the article page online and as indicated below:

LastName, A.A.; LastName, B.B.; LastName, C.C. Article Title. *Journal Name* **Year**, *Article Number*, Page Range.

ISBN 978-3-03928-180-0 (Pbk)
ISBN 978-3-03928-181-7 (PDF)

Contents

About the Special Issue Editors . . . . **vii**

Francesco Mancini and Riccardo Salvini
Applications of Photogrammetry for Environmental Research
Reprinted from: *ISPRS Int. J. Geo-Inf.* **2019**, *8*, 542, doi:10.3390/ijgi8120542 . . . . **1**

Doug Stead, Davide Donati, Andrea Wolter and Matthieu Sturzenegger
Application of Remote Sensing to the Investigation of Rock Slopes: Experience Gained and
Lessons Learned
Reprinted from: *ISPRS Int. J. Geo-Inf.* **2019**, *8*, 296, doi:10.3390/ijgi8070296 . . . . **4**

Carlo Robiati, Matt Eyre, Claudio Vanneschi, Mirko Francioni, Adam Venn and John Coggan
Application of Remote Sensing Data for Evaluation of Rockfall Potential within a Quarry Slope
Reprinted from: *ISPRS Int. J. Geo-Inf.* **2019**, *8*, 367, doi:10.3390/ijgi8090367 . . . . **28**

Claudio Vanneschi, Marco Di Camillo, Eros Aiello, Filippo Bonciani and Riccardo Salvini
SfM-MVS Photogrammetry for Rockfall Analysis and Hazard Assessment Along the Ancient
Roman Via Flaminia Road at the Furlo Gorge (Italy)
Reprinted from: *ISPRS Int. J. Geo-Inf.* **2019**, *8*, 325, doi:10.3390/ijgi8080325 . . . . **52**

**Rudolf Urban, Martin Štroner, Peter Blistan, Ľudovít Kovanič, Matej Patera, Stanislav Jacko,
Igor Ďuriška, Miroslav Kelemen and Stanislav Szabo**
The Suitability of UAS for Mass Movement Monitoring Caused by Torrential Rainfall—A Study
on the Talus Cones in the Alpine Terrain in High Tatras, Slovakia
Reprinted from: *ISPRS Int. J. Geo-Inf.* **2019**, *8*, 317, doi:10.3390/ijgi8080317 . . . . **75**

**Marion Jaud, Christophe Delacourt, Nicolas Le Dantec, Pascal Allemand, Jérôme Ammann,
Philippe Grandjean, Henri Nouaille, Christophe Prunier, Véronique Cuq,
Emmanuel Augereau, Lucie Cocquempot and France Floc'h**
Diachronic UAV Photogrammetry of a Sandy Beach in Brittany (France) for a Long-Term
Coastal Observatory
Reprinted from: *ISPRS Int. J. Geo-Inf.* **2019**, *8*, 267, doi:10.3390/ijgi8060267 . . . . **93**

**Konstantinos Nikolakopoulos, Aggeliki Kyriou, Ioannis Koukouvelas, Vasiliki Zygouri and
Dionysios Apostolopoulos**
Combination of Aerial, Satellite, and UAV Photogrammetry for Mapping the Diachronic
Coastline Evolution: The Case of Lefkada Island
Reprinted from: *ISPRS Int. J. Geo-Inf.* **2019**, *8*, 489, doi:10.3390/ijgi8110489 . . . . **106**

Mingbo Liu, Chunxiang Cao, Wei Chen and Xuejun Wang
Mapping Canopy Heights of Poplar Plantations in Plain Areas Using ZY3-02 Stereo
and Multispectral Data
Reprinted from: *ISPRS Int. J. Geo-Inf.* **2019**, *8*, 106, doi:10.3390/ijgi8030106 . . . . **132**

About the Special Issue Editors

Francesco Mancini received his M.Sc. degrees in Marine Environmental Sciences and a Ph.D. in Geodetic Sciences and Topography from the University of Bologna, Italy. Since 20 years, he works in the field of geomatics, as past researcher (University of Bologna and Technical University of Bari, Italy) and currently associate professor (University of Modena and Reggio Emilia, Italy). He is involved as surveyor in application to ground deformation, natural hazard assessment and landscape archaeology. He is professor in Geomatics technology, Geographical Information System, Cartography and Photogrammetry at schools of Civil and Environmental Engineering, Geology and Geography in diversified academic institutions. His current interests include, precision surveys, spatial data analysis, ground subsidence, monitoring of coastal areas, GNSS positioning and photogrammetric surveys by unmanned aerial vehicle (UAV). He has published a number of papers on international journals focused on surveying for environmental researches and modeling of natural/anthropogenic phenomena.

Riccardo Salvini received his M.Sc. and Ph. degrees in Geological Sciences from the University of Siena, Italy, respectively in 1996 and 2001. In 1998, he awarded with distinction the Professional Master degree in Geoinformatics (Photogrammetry and Remote Sensing specialization) at ITC (International Institute of Aerospace Survey and Earth Sciences, the Netherlands). Since 2003 he is Assistant Professor at the University of Siena and he mainly works in the fields of Remote Sensing, Digital Photogrammetry, GNSS positioning and Laser Scanning applied to geological investigations. He is involved in researches in Italy (Apuan Alps, Northern Apennines, Western Alps), Brazil (Mato Grosso and Mato Grosso do Sul), Libya (Cyrenaica), Turkey (Central Anatolia), Egypt (North-Eastern Sahara Desert), United Arab Emirates (Abu Dhabi), and Ethiopia (Rift Valley). He was involved in national and international projects on rock slope stability, soil erosion, multitemporal morphological and land use changes, and land subsidence. He is lecturer of Photogeology and Geotechnics for the B.Sc. degree in Geological Sciences, lecturer of Remote Sensing, Photogrammetry and Laser Scanning for the M.Sc. degree in Geomatics, and lecturer of Slope Stability for the M.Sc. degree in Engineering Geology. He is author and co-author of several publications in journals, papers in volumes and proceedings in conferences on the above-mentioned topics.

 International Journal of
Geo-Information

Editorial

Applications of Photogrammetry for Environmental Research

Francesco Mancini [1,*] and Riccardo Salvini [2]

[1] Department of Engineering 'Enzo Ferrari', University of Modena and Reggio Emilia, 41125 Modena, Italy
[2] Department of Environment, Earth and Physical Sciences and Centre of GeoTechnologies, University of Siena, 53100 Siena, Italy; riccardo.salvini@unisi.it
* Correspondence: francesco.mancini@unimore.it

Received: 25 November 2019; Accepted: 26 November 2019; Published: 28 November 2019

The applications of photogrammetry for environmental research benefits from the continuous and rapid evolution of sensors and methodologies in this field. The support of photogrammetric tools to a very wide range of research activities was previously confined to geomatic disciplines and the methodologies strictly based on terrestrial or traditional aerial photogrammetry. However, the timely investigation of natural or anthropogenic phenomena required more flexible tools and the ability of geoscientists and researchers involved in the study of natural resources to exploit photogrammetric methodologies in a more flexible way. In the last decade, new opportunities came from the possibility to acquire images using low-cost non-metric cameras from low-altitude unmanned aerial vehicles (UAVs), or fixed locations in terrestrial surveys, with a successive highly automated processing strategy. For instance, a huge amount of papers in the recent scientific literature refers to the structure from motion (SfM) technology in the reconstruction of three-dimensional features at very high spatial and temporal resolutions and with a surprisingly high positional accuracy. Point clouds obtained from best practices in novel approaches of close-range photogrammetry has proven to be of compatible spatial resolution and accuracy of those provided by terrestrial laser scanning and, very often, photogrammetry and laser scanning are combined to enhance the qualities of each other.

However, a limited number of papers were focused on ongoing processes, dynamic assessments or used in modeling of complex phenomena starting from single or repeated photogrammetric surveys with careful design of the timing of investigations. The Special Issue aimed for papers including novelties and advances on the use of recent photogrammetric approaches to a wide range of environmental studies, including the following: photogrammetry for monitoring; UAV photogrammetry for environmental research; photogrammetry for disaster prevention and management; photogrammetry for real-time mapping; merging of data from different survey technologies; and novel uses of proximity surveys to geography, geomorphology, geotechnologies, landscape description, coastal studies, archaeology, etc. Manuscripts on multitemporal investigation of environmental processes by the combined use of photogrammetry and other different technologies were also welcome for this Special Issue. After the revision procedure, seven papers strongly focused on the abovementioned topics have been selected and published. Two of them refer to applications on coastal monitoring using multitemporal images acquired from aerial, satellite and unmanned aerial vehicle platforms. Nikolakopoulos et al. [1] used Pleiades remote sensing data and aerial photogrammetry to quantify the historical rate of coastal erosion in the southwestern Lefkada (Ionian Sea, Greece) coastline. The paper by Jaud et al. [2] describes the activities of the monitoring of Porsmilin Beach (Brittany, France), carried out since 2006 and based on drone photogrammetry; storm impacts and beach resilience have been assessed by a long-term time series of UAV images acquired along a period characterized by multiple technological evolutions. Such a paper faced issues related to data quality/consistency and the need of high accuracy in the generation of digital elevation models (DEM) and orthophotos for coastal hazard purposes. Other papers focused on topics related to investigations on geomorphological hazard: Robiati et al. [3] proposed an application about the use of aerial LiDAR (Light Detection and Ranging) and photogrammetric campaigns to evaluate and back-analyze the rockfall potential, over almost a decade, in an active quarry located in Cornwall, UK. A methodology to characterize the orientation of discontinuities present within the rock slope is discussed and evidences for potential rockfall evolution were also addressed. The authors presented the use of aerial and terrestrial LiDAR data for the reconstruction of fine surface topography, rock slope kinematic analysis and rockfall trajectory modelling by using both 2D and 3D numerical simulations. Rockfall events have been investigated also by Vanneschi et al. [4]. This paper discusses the ability of the structure from motion (SfM) technique and multi view stereo (MVS) photogrammetry to perform rockfall analyses and hazard assessments. The case study is represented by the Ancient Roman Via Flaminia Road at the Furlo Gorge (Italy). In this paper, traditional geological methods of engineering geology have been combined with terrestrial laser scanning and drone-based digital photogrammetry for successive rock slope stability analyses and 3D rockfall runout simulations. Results show the rockfall hazard in the study area bring into evidence the fundamental role of detailed photogrammetric surface models in the reconstruction of slope, joints and block geometries. Moreover, this Special Issue includes a paper by Urban et al. [5] that deals with the prediction of landslides supported by UAV photogrammetry and laser scanning in mountainous environments (Malá Studená Dolina, Little Cold Valley, High Tatras National Park, Slovakia). The authors discuss logistic constraints, methodologies and accuracy achieved in the 3D reconstruction of talus cones in mountainous terrain hosting seasonal, heavy-used hiking trails. Finally, Liu et al. [6] present a paper on the application of satellite-based photogrammetry to map canopy heights of poplar plantations in plain areas (Horqin Sandy Land, eastern Inner Mongolia, China). In particular, the canopy heights have been mapped through a combination of stereo and multispectral data provided by China's latest civilian stereo mapping satellite ZY3-02.

In addition to such research papers, a review by Stead et al. [7] on the application of remote sensing to the investigation of rock slopes is included in the Special Issues. In such a review, the authors discuss a range of applications of field and remote sensing approaches for the characterization of rock slopes at various scales and distances and highlight advantages and limitations of the methodologies nowadays available in the accurate 3D representation of rock slopes.

The editors hope that the scientific community involved in photogrammetry and remote sensing for environmental applications will find the papers published in this Special Issue useful for their future investigations in a similar field.

References

1. Nikolakopoulos, K.; Kyriou, A.; Koukouvelas, I.; Zygouri, V.; Apostolopoulos, D. Combination of Aerial, Satellite, and UAV Photogrammetry for Mapping the Diachronic Coastline Evolution: The Case of Lefkada Island. *ISPRS Int. J. Geo-Inf.* **2019**, *8*, 489. [CrossRef]
2. Jaud, M.; Delacourt, C.; Le Dantec, N.; Allemand, P.; Ammann, J.; Grandjean, P.; Nouaille, H.; Prunier, C.; Cuq, V.; Augereau, E.; et al. Diachronic UAV Photogrammetry of a Sandy Beach in Brittany (France) for a Long-Term Coastal Observatory. *ISPRS Int. J. Geo-Inf.* **2019**, *8*, 267. [CrossRef]
3. Robiati, C.; Eyre, M.; Vanneschi, C.; Francioni, M.; Venn, A.; Coggan, J. Application of Remote Sensing Data for Evaluation of Rockfall Potential within a Quarry Slope. *ISPRS Int. J. Geo-Inf.* **2019**, *8*, 367. [CrossRef]
4. Vanneschi, C.; Di Camillo, M.; Aiello, E.; Bonciani, F.; Salvini, R. SfM-MVS Photogrammetry for Rockfall Analysis and Hazard Assessment Along the Ancient Roman Via Flaminia Road at the Furlo Gorge (Italy). *ISPRS Int. J. Geo-Inf.* **2019**, *8*, 325. [CrossRef]
5. Urban, R.; Štroner, M.; Blistan, P.; Kovanič, Ľ.; Patera, M.; Jacko, S.; Ďuriška, I.; Kelemen, M.; Szabo, S. The Suitability of UAS for Mass Movement Monitoring Caused by Torrential Rainfall—A Study on the Talus Cones in the Alpine Terrain in High Tatras, Slovakia. *ISPRS Int. J. Geo-Inf.* **2019**, *8*, 317. [CrossRef]

6.   Liu, M.; Cao, C.; Chen, W.; Wang, X. Mapping Canopy Heights of Poplar Plantations in Plain Areas Using ZY3-02 Stereo and Multispectral Data. *ISPRS Int. J. Geo-Inf.* **2019**, *8*, 106. [CrossRef]

7.   Stead, D.; Donati, D.; Wolter, A.; Sturzenegger, M. Application of Remote Sensing to the Investigation of Rock Slopes: Experience Gained and Lessons Learned. *ISPRS Int. J. Geo-Inf.* **2019**, *8*, 296. [CrossRef]

 International Journal of *Geo-Information*

Review

Application of Remote Sensing to the Investigation of Rock Slopes: Experience Gained and Lessons Learned

Doug Stead [1], Davide Donati [1,*], Andrea Wolter [2] and Matthieu Sturzenegger [3]

[1]   Department of Earth Sciences, Simon Fraser University, Burnaby, BC V5A 1S6, Canada
[2]   Palmer Environmental Consulting Group, Vancouver, BC V6C 1V5, Canada
[3]   BGC Engineering, Vancouver, BC V6Z 0C8, Canada
*   Correspondence: davide_donati@sfu.ca

Received: 10 May 2019; Accepted: 24 June 2019; Published: 27 June 2019

Abstract: The stability and deformation behavior of high rock slopes depends on many factors, including geological structures, lithology, geomorphic processes, stress distribution, and groundwater regime. A comprehensive mapping program is, therefore, required to investigate and assess the stability of high rock slopes. However, slope steepness, rockfalls and ongoing instability, difficult terrain, and other safety concerns may prevent the collection of data by means of traditional field techniques. Therefore, remote sensing methods are often critical to perform an effective investigation. In this paper, we describe the application of field and remote sensing approaches for the characterization of rock slopes at various scale and distances. Based on over 15 years of the experience gained by the Engineering Geology and Resource Geotechnics Research Group at Simon Fraser University (Vancouver, Canada), we provide a summary of the potential applications, advantages, and limitations of varied remote sensing techniques for comprehensive characterization of rock slopes. We illustrate how remote sensing methods have been critical in performing rock slope investigations. However, we observe that traditional field methods still remain indispensable to collect important intact rock and discontinuity condition data.

Keywords: Remote sensing; field work; slope stability; landslide mapping; damage

1. Introduction

Landslides are among the most destructive natural phenomena and are responsible for several hundreds of deaths every year. Continuous population growth has resulted in the progressive settlement of steeper slopes, causing more people to live and work in areas with high landslide risk [1]. Additionally, the increased likelihood of extreme weather events associated with climate change tendencies can trigger major landslides [2,3].

The development of rock slope instabilities is controlled by many factors. Structural and lithological features, such as faults, folds, joints, foliation, and bedding planes, can provide kinematic freedom to potentially unstable blocks at the scale of outcrops and roadcuts to entire mountain slope [4,5]. Brittle fracturing of intact rock can contribute to the evolution of rock slope instability by allowing the formation of continuous failure surfaces [6,7]. The instability of rock slopes is also enhanced by endogenic and exogenic factors, such as earthquakes [8], groundwater fluctuations [9], glacial retreat [10], and slope erosion and steepening [11]. It is clear that a comprehensive rock slope characterization is required to identify the factors controlling stability, and to investigate how their potential impacts vary both spatially and temporally [12].

Traditional field techniques allow geoscientists and engineers to systematically collect important geomechanical parameters for both intact rock and discontinuities that can be employed in preliminary and advanced stability analyses. However, data collection is spatially restricted to accessible areas and may be limited or prevented by difficult terrain, active instability, steep slopes, and/or time and resource constraints. The development and technological advancements of remote sensing techniques have allowed geological data to be effectively collected across inaccessible slopes [13]. Photogrammetric techniques, such as terrestrial digital photogrammetry (TDP) and structure-from-motion (SfM), are routinely used to perform rock mass characterization [13–15] and geomorphic mapping [16]. Airborne and terrestrial laser scanning (ALS and TLS) provide 3D point clouds that can be employed to map geological structures [17] and monitor displacements [18]. Recently, infrared techniques, such as infrared thermography (IRT) and hyperspectral imagery (HSI), have been introduced to investigate seepage and rock slope mineralogy, respectively [19,20]. The integrated application of multiple remote sensing techniques using an integrated approach may improve understanding of the mechanisms underlying slope instability. Presently, multisensor remote sensing applications predominantly involve the coupling of TLS datasets with IRT [21] or HSI datasets [22].

In this paper, we describe and summarize the methods and procedures used for data collection and rock mass and slope characterization at various sites, both in Canada and in Italy (Figure 1), investigated during the past 15 years by current and former members of the Engineering Geology and Resource Geotechnics Research Group at Simon Fraser University. The sites include the Hope Slide and Block 731 (British Columbia, Canada), the Frank Slide and the Palliser rockslide (Alberta, Canada), and the Vajont Slide (Italy). In the paper we review, for each site, the methods used and results obtained from traditional field characterization, and describe how knowledge of the site and instability processes progressively increased through collection and interpretation of remote sensing datasets. For each slope, we present examples of collected datasets, highlighting the potential application and advantages of each remote sensing technique in the characterization of rock mass and slope damage.

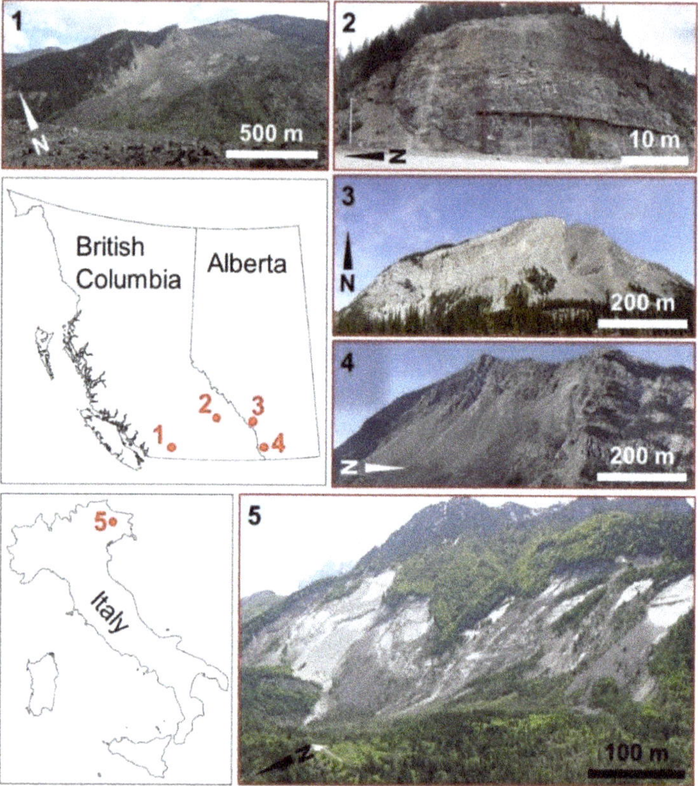

Figure 1. Location of the sites described in this paper. **1**: Hope Slide (British Columbia); **2**: Block 731 (British Columbia); **3**: Palliser rockslide (Alberta); **4**: Frank Slide (Alberta); **5**: Vajont Slide (Italy).

2. Methodology

The comprehensive investigation of rock slopes and landslides is a complex task that comprises (a) review and analysis of literature and historical data, (b) data collection by means of both traditional field techniques and remote sensing methods, and (c) data analysis and numerical modelling [12]. In this paper, we describe field and remote sensing techniques and procedures for the collection of geotechnical data at the selected rock slopes.

Traditional field-based mapping techniques were used for the characterization of the rock mass and the systematic collection of discontinuity data using scanline and/or window mapping. The uniaxial compressive strength (UCS) of intact rock was estimated in-situ using a geological hammer or the Schmidt hammer. Intact rock specimens were collected for point load testing (PLT), which provides more reliable uniaxial strength estimations [23]. The International Society for Rock Mechanics and Rock Engineering provides guidelines for the collection of discontinuity data at the rock face [24]. Water conditions, surface roughness, and type of infill were characterized to determine the shear strength of the discontinuities [25]. Rock mass blockiness and block shape were determined based on discontinuity orientation, persistence, and spacing. Rock mass quality was estimated using the Geological Strength Index (GSI), based on the configuration and surface conditions of the discontinuities. GSI was also used to estimate rock mass strength and deformability [26]. Field activity was also useful for the geomorphological characterization of landslide deposits, particularly if high-resolution aerial datasets were not readily available [27,28].

To complement traditional mapping techniques, various remote sensing techniques were used for the characterization of the rock slopes, including high-resolution photography and digital photogrammetric methods (TDP and SfM), ALS and TLS, IRT, and HSI. A brief overview of each method and the survey equipment employed is provided.

2.1. Digital Photogrammetry and High-Resolution Photographs

Digital photogrammetric techniques, such as TDP and SfM, allow 3D models of the investigated slopes to be constructed using photographs taken from at least two different locations [29,30]. The scene reconstruction is performed using a semi-automated procedure that allows identification and matching of pixels (in TDP) or features (in SfM) within the photographs and using a bundle-adjustment approach to estimate their location in 3D. In this study, TDP was largely employed to perform discontinuity mapping and rock mass characterization on the 3D slope models. SfM was employed to perform a block size analysis of the rock avalanche deposit of the Hope Slide, using photographs collected with an unmanned aerial vehicle (UAV). Photographs at the investigated sites were collected with several digital cameras for TDP, including Canon EOS 30D (8.2 MegaPixel), Canon EOS 5D mark II (21.1 MegaPixel), and Canon EOS 5Ds-R (50.6 MegaPixel, Figure 2a). SfM photographs were capture using a DJI Phantom 3 Pro Quadcopter with an integrated 20 MegaPixel camera (Figure 2b). Processing was performed using the 3DM Analyst mapping suite 2.5 [31] for TDP and Photoscan 1.4 [32] for SfM.

The collection of high-resolution photographs is useful for the identification and mapping of slope damage features, such as open cracks, areas of surface weathering, and presence of discontinuity infill. Characterization of these parameters is difficult on lower resolution TDP and SfM models and, therefore, required high-resolution photographs.

2.2. TLS and ALS

Laser scanning techniques allow 3D point clouds of the investigated slopes to be obtained. In the point clouds, the location of each point is estimated based on the direction and time-of-flight (ToF) of a laser pulse emitted from the instrument and reflected by the investigated slopes [17]. Processing of ALS datasets allows vegetation to be digitally removed, and bare earth (BE) datasets to be created. BE datasets are mainly employed for large-scale structural analysis, geomorphic mapping, and monitoring [13]. However, due to the vertical line of sight and lower resolution, they are not suited for the investigation of steep slopes. Conversely, due to their oblique line of sight and higher resolution, TLS are routinely employed for rock mass characterization, discontinuity mapping, and small-scale slope monitoring [18,33]. In this paper, TLS datasets were collected using Optech ILRIS3D (1500 m max range, Figure 2c) and Riegl VZ-4000 (4000 m max range, Figure 2d) terrestrial laser scanners. Collected point clouds were processed in RiScan Pro 2.6 [34], CloudCompare 2.9 [35], and Polyworks [36].

2.3. IRT

IRT surveys allow the temperature distribution of an investigated object to be examined. Any object with a temperature higher than 0 K is a source of infrared energy. Since the infrared emission is a function of temperature, a thermal camera can capture and convert the infrared radiation into a temperature map of the object [37]. In this paper, IRT was conducted predominantly to investigate and map discontinuity seepage using a FLIR SC7750 thermal camera (Figure 2e). IRT imagery was processed using ResearchIR [38].

2.4. HSI

HSI is a technique that allows the electromagnetic reflected radiation of an object to be investigated. Infrared radiation in the short-wave infrared (SWIR) spectrum is diagnostic of the mineralogical components of natural materials and can be used to identify lithological contacts and variations [39]. HSI is largely employed in the field of mineral exploration, using satellite-based imagery [40]. The recent introduction of ground-based HSI push-broom scanners allows the rock slope lithologies to be mapped using portable hyperspectral cameras [41]. In this paper, we employed a Specim SWIR3 hyperspectral scanner to identify lithological variations at both the outcrop- and mountain-slope scales (Figure 2f). Data processing has been undertaken using the software ENVI 5.5 [42].

Figure 2. Remote sensing equipment employed at the investigated sites. **a**: Canon EOS 5Ds-R with f = 400 telephoto lens; **b**: DJI Phantom 3 Pro Quadcopter; **c**: Optech ILRIS3D terrestrial laser scanning (TLS); **d**: Riegl VZ-4000 TLS; **e**: FLIR SC7750 thermal camera; **f**: Specim SWIR3 hyperspectral scanner.

3. Rock Slope and Landslide Investigations

3.1. Hope Slide (British Columbia, Canada)

The Hope Slide was a major rock avalanche that occurred on January 9th, 1965. The 48 million m^3 slide detached from the southern slope of Johnson Ridge, along BC Highway 3, 15 km east of the town of Hope. The rock avalanche travelled for two kilometers along the Nicolum Valley, infilling the valley floor with about 70 m of material, burying Outram Lake and causing the death of four people who were travelling along the highway at the time of failure [43]. The slide occurred in two stages, separated by a few hours. Two low-magnitude earthquakes were recorded at the site, which are suggested to have resulted from the slope failures [44]. From a geological perspective, the failed slope is composed of a massive greenstone of the Hozameen Group, which is locally intruded by felsite sills and dikes [45]. Various north-south striking faults were observed to cross the slide area, which were suggested to have controlled structurally the location and behavior of the slope instability [45]. The lower part of the slope was affected in prehistoric times by another rock avalanche, roughly similar in size, that likely occurred as a result of glacial retreat [46].

3.1.1. Field Work

Extensive field work at the Hope Slide site was conducted between 2003 and 2004 [47]. Traditional field work techniques were employed to perform rock mass characterization and estimate the spatial variation of the rock mass quality using the GSI. Field work investigations were conducted along the lower part of the lateral scarp, at the base of the rupture surface, and in the upper slope. Engineering geological mapping allowed four structural domains to be identified, based on the orientation of seven observed discontinuity sets. Discontinuity set J1 was found to be sub-parallel to the slope and was suggested to have acted as the sliding surface for the 1965 event. Faults and shear zones identified within the slide area were also investigated and weak gouge up to 30 cm thick was observed at their core. Unconfined compressive strength of the gouge was estimated between 12.5 and 50 MPa, while the intact strength of the greenstone was estimated at 159 MPa using point load testing [47]. The lowest rock mass quality was observed along the lateral scarp, which comprise a shear zone with GSI values as low as 0–10. GSI was found to vary between 20–30 and 70–80 in the central part of the slide area, and between 10–20 and 50–60 in the southern part of the upper slope [47].

From the field-based slope characterization, it was observed that large-scale geological structures were responsible for a high variability of the observed joint sets. Additionally, it was suggested that faults and shear zones divide the failed slope into different blocks, highlighting the strong 3D structural control that governed the 1965 event [47].

3.1.2. Remote Sensing

Extensive remote sensing was conducted at the site in summer 2011 and summer 2015, using SfM, TDP, TLS, and IRT. Results are described in [12,46,48]. Historic aerial photographs, collected in 1961 (i.e., prior to the 1965 rock avalanche events) and retrieved from the Province of British Columbia database, were used to build a pre-failure 3D model using SfM. The 3D model was investigated in ArcGIS 10.5 [49], and a large-scale structural analysis was undertaken using hillshade, slope, and aspect maps. Six major lineaments were identified, which divide the slope into five blocks. Two of these blocks were likely involved in the prehistoric slope failure (Figure 3a).

A TDP survey was performed from two stations located within the debris field, using a Canon EOS 5D mark II and an f = 400 mm focal length lens. Discontinuity mapping was performed on the constructed 3D models, and over 1600 discontinuity orientations were measured from the inaccessible parts of both the headscarp and sliding surface (Figure 3b). Three discontinuity sets were identified, consistent with the major discontinuity sets mapped using traditional field techniques. J1 was found to be sub-parallel to the slope, while J2 and J3 are perpendicular to J1 and to one another. Based on the discontinuity orientations, five structural domains were identified, characterized by a progressive rotation of the observed discontinuity sets that was not clearly recognized in previous field-based analyses [48] (Figure 3c).

An aerial photogrammetric survey was undertaken using a UAV to investigate block size distribution within the slide debris. About 650 photographs were collected and used to build an orthorectified photograph in Photoscan (Figure 3d). 2000 blocks were then manually digitized, and their volume estimated. The largest block was estimated at 4000 m^3 whereas the average block size was about 78 m^3.

A TLS survey was conducted from the Hope Slide visitor lookout at the base of the slope in the summer 2015. The elevation difference between the point cloud and the pre-failure model built using historical aerial photographs was computed in ArcGIS to estimate the volume involved in the failure. A volume of 48.4 million m^3 was calculated, which is in good agreement with previous estimations, ranging between 47.3 and 48.3 million m^3, that were based on pre- and post-failure isopach maps [43,45].

An IRT survey was undertaken to investigate seepage along the daylighting rupture surface, and it was observed that most of the seepage occurs along the discontinuity set J1 (Figure 3e).

Traditional field work and remote sensing data were employed to create a 3D numerical model of the slope and to simulate the failure of the Hope Slide through progressive cohesion reduction along the basal rupture surface using a 3D distinct element modelling approach [12]. The numerical analysis allowed a two-stage failure to be reproduced, which was probably caused by the varied kinematic conditions across the unstable slope.

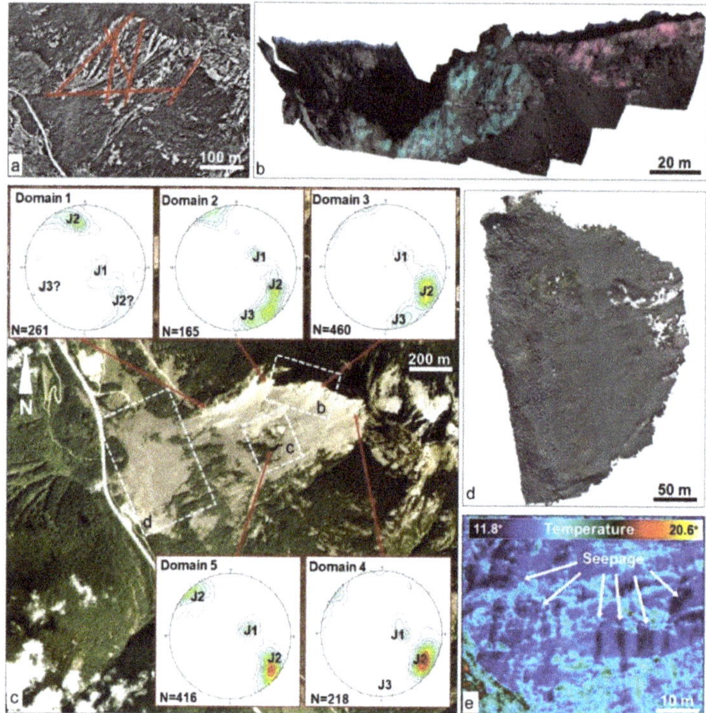

Figure 3. Remote sensing analysis performed at the Hope Slide. **a**: pre-failure 3D model of the slope, constructed using 1961 aerial imagery processed using the structure-from-motion (SfM) technique; **b**: example of a 3D model reconstructed using terrestrial digital photogrammetry (TDP). Blue and pink disks show discontinuities mapped in 3DM Analyst; **c**: structural overview of the area. The stereonets show the orientation of the discontinuity sets mapped in the TDP models, subdivided into five structural domains; **d**: orthorectified image of the debris field reconstructed using unmanned aerial vehicle (UAV)-SfM; **e**: infrared thermography (IRT) dataset of the daylighting sliding surface. Dark areas identify seepage.

3.2. Block 731 (British Columbia, Canada)

Block 731 refers to a stabilized, 55 m high rock slope located along the left abutment of the Revelstoke Dam, in British Columbia, Canada [50]. The Revelstoke Dam comprises a 160 m-high concrete gravity dam, a powerhouse located immediately beneath the concrete dam, and an earthfill dam that extends for 1 km to the west of the concrete structure, over the west bank terrace of the Columbia River [51]. During the construction of the dam (1977–1984), a 280,000 m^3 rock slope instability developed along the eastern abutment. The instability extended over a length of ca. 200 m and was triggered by blasting for the excavation of the Highway 23 bypass road cut. No attempt was undertaken to stabilize the southernmost part of the unstable slope, referred to as December Slide (30,000 m^3), which was progressively excavated, and the material removed. The northern part of the unstable area, referred to as Block 731, involved displacement of a volume of 250,000 m^3 of rock mass along a weak shear zone, identified as S3 [50]. Displacement rates up to 3.6 mm/day were recorded during the excavation of the road cut. The installation of a temporary beam at the base of the slope reduced the displacement rates to 0.2 mm/day, allowing permanent remedial measures, comprising anchors and drainage systems, to be installed [50].

The rock mass forming the Block 731 predominantly comprises an alternation of quartzite and horneblend gneiss [50]. These lithologies are part of the Selkirk Allochthone, and specifically the tectonic slice referred to as Clachnacudainn [52]. The Selkirk Allochthone forms the hanging wall of the Columbia River Fault, which is the most important geological structure in the area [53].

3.2.1. Field Work

Field work at the site was undertaken in summer 2017 and involved preliminary discontinuity mapping and rock mass quality estimation using the GSI. A visual analysis of the rock slope, characterized by a dip/dip direction of 63°/296°, showed the presence of two major shear zones, located respectively at the northern edge (3–4 m wide) and in the southern part of the rock slope (1–2 m wide). These appear to correspond to the shear zones S3 and S4, respectively [50].

Preliminary discontinuity mapping undertaken at the site showed the presence of four main discontinuity sets, including the foliation. J1 and J2 dip into the slope with orientations (dip/dip direction) of 65°/128° and 73°/092°, respectively. J3 is a high angle discontinuity set perpendicular to the slope orientation (83°/009°). The foliation in the lower part of the slope dips into the slope at a low angle (36°/120°). It was noted that foliation planes are characterized by smoother surfaces (JCS = 4–10) compared to discontinuity sets J1–J3 (JCS = 8–12).

The rock mass in the shear zones is highly altered, weathered, and fractured, and is easily excavated using a geological hammer (GSI = 0–10). Shotcrete was sprayed to prevent material detachment from the shear zones, but large areas of shotcrete were missing, particularly along the lower slope, at the time of the investigation. Outside of the shear zones, rock mass structure appears very blocky, and overall rock mass quality increases (GSI = 45–55).

3.2.2. Remote Sensing

Remote sensing surveys were conducted in the summer of 2017 and including high-resolution photography, TLS, and HSI analyses. High-resolution photography was collected to produce a detailed panoramic image of the investigated slope, and to complement discontinuity mapping completed on the TLS dataset (Figure 4a). In particular, high-resolution photography was employed to distinguish between tectonic discontinuities, blast damage features, and planar surfaces.

The TLS dataset was collected using the Riegl VZ-4000 from two scan positions located at the base of the slope, on the opposite side of Highway 23. A point spacing ranging between 2 and 4 cm was obtained for the collected point cloud (Figure 4b). Four discontinuity sets were identified, including the foliation, consistent with traditional field-based discontinuity mapping, and TLS structural mapping. The average orientation of the foliation and sets J1 and J2 were estimated as 28°/119°, 63°/138°, and 64°/097°, respectively. The orientation of J3 was found to differ between traditional and TLS mapping, due to the higher spatial coverage and number of features mapped on the TLS point cloud. The orientations of the shear zones S3 and S4 were estimated at 48°/332° and 41°/273°, respectively, using the TLS dataset. The analysis described here highlighted the presence a J2, which was not documented in previous analyses [54].

A preliminary HSI analysis was conducted, aiming to investigate the lithological variations that can be observed in the lower part of the slope. The imagery was collected using the SWIR3 hyperspectral scanner from the opposite side of Highway 23. At a distance of 10 m from the slope, the average ground pixel size is ca. 5–10 mm. The collected imagery was processed using an approach similar to [41] and comprised image de-striping, spectral filtering, minimum noise fraction (MNF) transformation [55], and empirical line correction [56]. The processed imagery allows the boundaries between different lithologies to be highlighted. In particular, an alternation between two different geological materials was observed. Locally, augens up to 3 m wide were also highlighted (Figure 4c).

Figure 4. Multi-sensor remote sensing analysis of the Block 731 road cut on the left abutment of the Revelstoke Dam. **a**: Panoramic overview of the 55 m high rock slope. Note the location and orientation of the shear zones S3 and S4; **b**: point cloud of the rock slope. The yellow box in **a** and **b** shows the location of sections in **c** and **d**; **c**: minimum noise fraction (MNF) image of part of the lower slope comprising the three lowest noise fraction bands obtained from the hyperspectral imagery (HSI) dataset. Details of augens and lithological alternations are shown. Note the high contrast between different geological materials; **d**: high-resolution photograph of the same part of the slope shown in **c**.

3.3. Palliser Rockslide (Alberta, Canada)

The Palliser Rockslide is a major prehistoric slope failure located in the Canadian Rocky Mountains, 70 km south of Banff, Alberta. Two major failure events involved rock mass detachments from the western slope of Mt. Indefatigable, which constitutes the eastern limb of the Serrail Creek Syncline. The bedrock in the area comprises carbonate rocks of Devonian and Mississippian age [57]. It was noted that the rockslide deposit resulted from two distinct events, possibly separated by thousands of years, which, respectively, involved rock from the Livingstone Formation and the Upper Banff Formation [58]. Cosmogenic dating was performed using ^{36}Cl, noting that the slope failures occurred ca. 10,000 and ca. 7700 years ago, respectively [59]. The displacement occurred in a south-western direction along pervasive bedding planes sub-parallel to the slope orientation, separated by steps as described in Section 3.3.2. A maximum dip angle of 50° is observed at the top of the rupture surface, and progressively decreases towards the base of the slope approaching the axis of the large-scale syncline [60]. The north–western dipping limb of the Serrail Creek Syncline, at the base of the slope, was suggested to provide stability to the present-day slope, north of the prehistoric failure, by providing a "buttress" in which bedding planes dip into the slope [61].

3.3.1. Field work

Traditional field work was undertaken at the site in 2006 and 2008. A detailed description is provided in [61] and [62]. Rock mass characterization was performed, including discontinuity mapping, rock mass quality estimation, and a qualitative, preliminary groundwater analysis. Scanline mapping, undertaken at the base of the slope, revealed the presence of low to medium persistence, extremely close to closely spaced bedding planes (S0), and three discontinuity sets (J1, J2, and J3) characterized by low persistence and close spacing. Mapping conducted along the rupture surface showed the dip angle of the bedding planes to vary between 24° and 52°, with a progressive increase towards the top of the slope, due to the curvature of the north-western limb of the syncline [62]. Rock mass quality, estimated using GSI, was noted to vary between 55–65 and 65–75 in the area where the scanline mapping was undertaken. The rock mass within the Serrail Creek Syncline, daylighting along the northwestern boundary of the slide area, is characterized by a GSI of 30–40, due to a higher fracture intensity [62]. Locally, carbonate infill was identified along bedding planes and J1 discontinuities. Additionally, seepage, karst features, and carbonate recrystallization features were observed within the rock mass forming the lateral boundary of the slide, suggesting that groundwater circulation occurred within the slope, and may have played a role in the failure [62]. Intact rock samples were collected during the field work to estimate the UCS of the intact rock in directions parallel and perpendicular to the bedding; average values of 180 and 118 MPa, respectively, were obtained from point load testing on thirty samples [62].

3.3.2. Remote Sensing

Remote sensing investigations were conducted at the site in 2006, 2008 [61,62], and 2017. In 2006 and 2008, TDP surveys were conducted from multiple locations within the deposit using a Canon EOS 30D with an f = 20–400 mm focal length lens. The 3D models reconstructed using an f = 50 mm focal length lens were processed, based on geomorphic field evidences, to estimate the total volume involved in slope failures. It was suggested that the volume of the younger event did not exceed 8 million m^3, whereas the older event caused the failure of 40 million m^3 of rock [62]. These estimations are significantly lower than previous analyses [58].

TDP models constructed using an f = 200 mm focal length lens were used to perform long-range discontinuity mapping, step-path surface analysis, intact rock fracture mapping, and bedding waviness estimation. Discontinuity mapping was performed separately for the slide area and the south-western limb of the Serrail Creek Syncline. Results showed good agreement with discontinuity field mapping. Step-path analysis conducted using the TDP 3D models allowed four types of step-path surfaces to be distinguished, based on the persistence, spacing, and orientation of the discontinuities forming the rupture surface [62]. Intact rock fracture mapping was conducted to estimate the percentage of rock bridges, which was noted to be 2%–3% along the failure surface. The waviness of bedding planes in the TDP models was investigated at multiple locations along the rupture surface. The roughness angle i was estimated at multiple locations and noted to vary between 4° and 5° [62].

In summer 2017, a multi-sensor remote sensing survey was undertaken at the site, involving the collection of TLS, high-resolution photography, and IRT datasets.

Two TLS surveys from distinct locations within the rockslide deposit were performed to build a high-density point cloud of the rock slope. The resulting point spacing is 5–7 cm from a distance of 1100 m (Figure 5b).

High-resolution photography was collected using the Canon 5Ds-R 50 MegaPixel camera with an f = 200 mm prime focal length lens. Photographs were collected from a single location within the deposit, at a distance of 1100 m from the top of the rupture surface, and are characterized by a ground pixel size of 2.4 cm. Draping high-resolution photographs onto 3D datasets was noted to enhance the detail of the 3D model [63], potentially allowing location, intensity, and distribution of intact rock fracturing features to be examined (Figure 5c).

The preliminary analysis of the TLS and high-resolution photographic datasets shows the presence of surface alteration along the sub-vertical rock bounding the slide area. Altered surfaces occur along the directions of maximum slope steepness, appearing as darkened surfaces in the high-resolution photographs (Figure 5d,g), and as high reflectance surfaces in the TLS dataset (Figure 5e,h). An IRT survey was conducted to investigate the presence of seepage along the rock slope, which did not show significant water flow in the areas where weathering was observed (Figure 5f,i). It appears that the surface alteration is probably caused by seepage or concentrated water flow occurring during rainfall events and snowmelt periods. Thermal anomalies along the rock slope are predominantly observed where karst features and shadows occur along the rock slope surface.

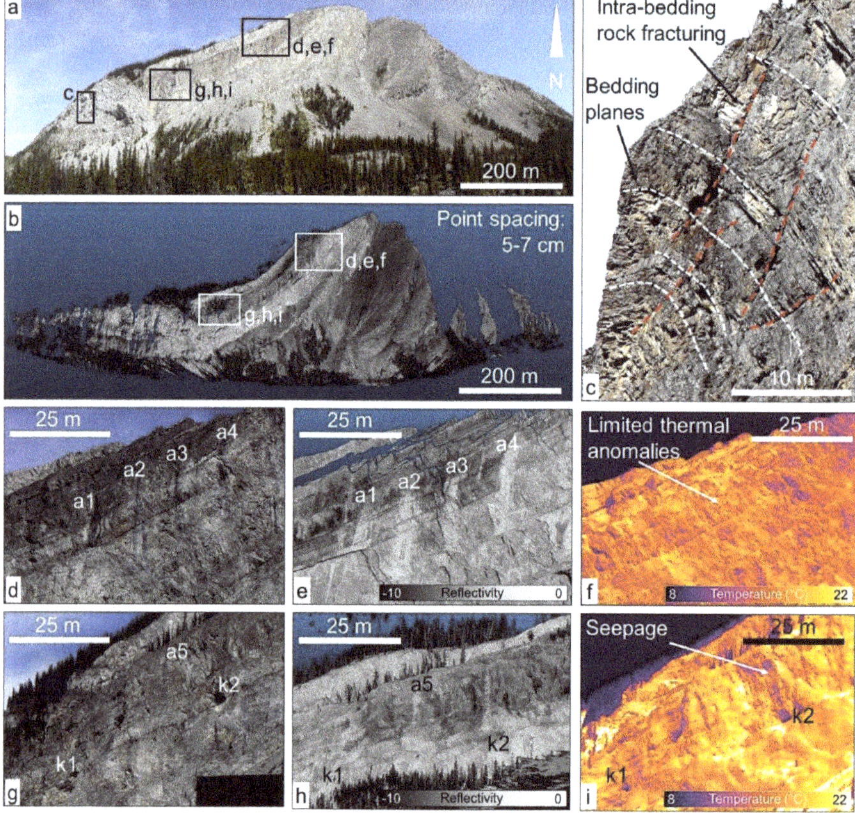

Figure 5. Remote sensing analysis performed at the Palliser Rockslide. **a:** panoramic photograph of the slope; **b:** TLS 3D point cloud; **c:** high-resolution photograph showing evidence of intra-bedding intact rock fracturing; **d–f:** details from the high-resolution photography, TLS, and IRT datasets showing areas of surface alteration (a1–a4) due to ephemeral seepage or concentrated water flow; **g–i:** details from the same datasets showing areas of surface alteration (a5) and karst features (k1, k2).

3.4. Frank Slide (Alberta, Canada)

The Frank Slide is a major rock avalanche (30 million m^3) that detached in 1903 from the eastern slope of Turtle Mountain, near the town of Frank [64]. The rock slide displaced along the eastern limb of Turtle Mountain anticline, which strikes in a north-south direction in the slide area [65]. The upper part of the slope, from which the slide detached, comprises limestone and dolostone of the Livingstone, Banff, and Palliser Formations, ranging in age from the Mississippian to Devonian. These formations thrusted over the shales and sandstone of the Fernie Formation and Kootenay Group, ranging in age from Jurassic to Cretaceous [65]. The Kootenay Group also comprises coal deposits, which were being mined at the time of the failure [66]. It has been suggested that coal mining played a role in weakening the slope, however, this hypothesis is not conclusive [64,67]. During the failure, tension cracks opened behind the headscarp, which have not shown further opening since the failure [65].

Presently, slope instability predominantly involves a volume of 7 million m^3 in the southern part of the headscarp, referred to as South Peak [68]. Monitoring of the area has been performed using crack meters, trilateral signs, photogrammetry, total station, and dGPS [69]. Since 2014, monitoring is largely based on ground-based InSAR (Interferometric synthetic-aperture radar installed at the base of the slope, which now represents the primary monitoring system at the site [70].

3.4.1. Field Work

Field work activity conducted along the potentially unstable South Peak area is described in [71]. Between 2008 and 2009, structural analysis and rock mass characterization were undertaken at the site. Discontinuity mapping, performed at various locations, allowed five major (J1–J5) and one minor (J6) discontinuity sets to be identified. The mapped discontinuities were noted to be planar or undulating (primary roughness) and characterized by rough surfaces (secondary roughness). Based on the varied orientation of the sets, the South Peak area was subdivided into three different structural domains. Rock mass quality was also estimated using a GSI approach and was observed to vary between 30–40 and 60–70 [71].

3.4.2. Remote Sensing

Various remote sensing datasets have been collected at the site, using a combination of TLS, TDP, high-resolution photography, IRT, and HSI. Preliminary TLS datasets, collected from five stations located at the North Peak and South Peak, were merged with TDP datasets to create the first composite 3D model of the headscarp [72]. TLS point clouds were obtained using the ILRIS3D TLS, whereas an f = 400 mm focal length lens was used to create the TDP models from the base of the slope, at a range of 2.1 km [72]. Structural mapping on the virtual outcrop were noted to agree with field measurements and allowed estimation of discontinuity persistence.

In summer 2017, a remote sensing survey campaign was undertaken at the site, aiming to collect multi-sensor datasets from very-long range (up to 2.8 km). TLS datasets were collected, using the VZ-4000 TLS, from three locations near the Frank Slide Interpretive Center and within the debris field. A point cloud of the headscarp was obtained, characterized by an average point spacing of 5–7 cm, thus potentially allowing for discontinuity mapping from a distance exceeding the range of the ILRIS3D TLS (Figure 6b). High-resolution photography, collected from the interpretive center with the Canon 5Ds-R and an f = 200 mm prime focal length lens, provided imagery characterized by a ground pixel size up to 5.8 cm. It was noted that despite the good resolution of the TLS dataset, the detailed structural investigation becomes challenging when smaller features need to be identified and mapped. Draping the high-resolution photographs onto a mesh obtained from the TLS dataset was found to be beneficial in identifying small-scale features in point clouds collected at long range (Figure 6c). The analysis of high-resolution photography also potentially allows rock bridges and brittle fractures along the rock slope to be examined.

IRT and HSI datasets were also collected from the interpretive center. The ground pixel sizes obtained with the SC7790 thermal camera and the SWIR3 hyperspectral scanner are 84 cm and 120 cm, respectively. Thermal imagery is generally employed to investigate seepage. At the Frank Slide, temperature distribution throughout the rock slope was noted to be related to the varying exposure across the headscarp (Figure 6d). No appreciable groundwater seepage was observed, possibly due to the lack of rainfall events in the weeks prior to the survey. The collected HSI dataset allows irradiance variations between layers to be observed along the headscarp, possibly associated with varied degrees of dolomitization within the Mississippian Livingston Formation (Figure 6e). Vegetated areas are also clearly visible in the collected dataset.

Figure 6. Multi-sensor remote sensing imagery of the Frank Slide headscarp. **a**: panoramic overview; **b**: point cloud of the South Peak area. Red and yellow boxes outline the sections shown in **b–e**, respectively; **c**: High-resolution photograph draped onto the TLS dataset; **d**: IRT image of the Frank Slide. Low temperature areas are associated to surface topography variation; **e**: HSI image of the Frank Slide (Red band: 1047.0 nm; green band: 1601.9 nm; blue band: 2103.2 nm). Variations in irradiance show possible mineralogical changes between layers.

3.5. Vajont Slide (Italy)

The Vajont Slide is located in northern Italy, at the boundary between the Veneto and Friuli Venezia-Giulia regions, 2 km east of Longarone, a town in the Piave River Valley. This failure was a direct consequence of the construction of a hydroelectric dam along the Vajont gorge, and the subsequent impoundment of a reservoir that reactivated a prehistoric landslide on the northern flank of Mount Toc [73]. Due to the repeated lowering and raising of the reservoir level, and following a heavy rainfall period, on October 9th, 1963, the slope failed [74]. A volume of 250 million m^3 of rock slid into the reservoir, generating a displacement wave that completely destroyed the town of Longarone and other small villages along the Piave River Valley, after travelling through the Vajont gorge at high speed. Almost 2000 people died during the event.

The Vajont gorge comprises carbonate rocks of the Soccher Formation, Fonzaso Formation, and Calcare del Vajont Formation, ranging in age from Cretaceous to early Jurassic [73]. The rupture surface of the Vajont Slide occurs within the Fonzaso Formation, which comprises moderately thin to moderately thick micritic limestone beds, characterized by clayey interlayers in the upper part [75]. The displacement of the 1963 slide probably occurred along such clay layers, which were characterized by residual strength values, due to the shearing or cumulateve damage caused by the paleo-deformation [76]. The size and characteristics of the Vajont Slide were greatly affected by geological structures present in the area. The rupture surface was characterized by a chair-like shape, structurally controlled by the Erto Syncline [77]. The lateral and rear release surfaces of the slide were defined by the Col Tramontin Fault and the Col delle Erghene Fault, respectively [78]. The slide body was divided into two major blocks by the Massalezza Creek, which crossed the slide area in a north-south direction. It is suggested that the larger, western block moved first, followed by the eastern block [79].

3.5.1. Field Work

Field work activity was conducted in summer 2010 and 2011. A detailed description of the investigations and results can be found in [78] and [79]. Field investigation involved an in-depth geomorphic characterization of the slide deposit, and a preliminary structural analysis, aimed to collect discontinuity orientation and fold axis data.

Geomorphic mapping was undertaken along north-south oriented foot transects which were spaced at 50 m, throughout the entire slide deposit area. During the survey, geomorphic features (such as gullies and ridges) and variation in slope angle were recorded as data points and were later employed to create a detailed geomorphological map. Based on the distribution of ridges within the deposit, areas of compression and extension, developed during the 1963 slope failure, were interpreted [78].

3.5.2. Remote Sensing

A comprehensive remote sensing analysis was conducted to investigate the pre- and post-failure morphology of the slope, and the structural setting of the sliding surface. A comprehensive remote sensing investigation, including photogrammetric and laser scanning techniques were critical to perform a geomorphic analysis of the pre- and post-failure slope topography, and to investigate the structural geology of the rupture surface, which is today largely inaccessible. The pre-failure slope investigation was conducted based on the stereo-interpretation of 1960 historical, aerial photographs (Figure 7a). A pre-failure geomorphological map was created, which highlighted the location and extension of geomorphic benches. These benches are characterized by low slope angle (<25°), the largest being the Pian della Pozza bench, gently dipping into the slope. Multiple steep scarps were also identified throughout the unstable rock mass. In the upper slope, three major scarps were noted to delineate the M-shaped tension crack that formed in 1960. Below the Pian della Pozza bench, another steep scarp was observed, and was suggested to be related to the incipient failure that would occur in November 1960 [78]. Minor, shallow slope instabilities (i.e., debris slide scars) were also observed both at the base of the slope, near the reservoir, and in the upper slope, close to the intersection of the Massalezza Creek with the rear slide boundary. Additionally, hummocks, ridges, and depressions could be observed, evidencing the unstable nature of the slope [78]. The post-failure geomorphic analysis was conducted using 1963 historic air-photographs and an ALS dataset. Ridges and extensional features were mapped and interpreted to infer the direction of displacement of the blocks forming the slide.

A high-resolution ALS dataset was used to create aspect and slope maps in ArcGIS to perform a preliminary geomorphic analysis of the slide area and validate the features observed during the traditional geomorphological mapping.

A TDP analysis was conducted to investigate the structural setting of the exposed rupture surface [79]. Long range surveys were conducted, using a Canon EOS 5D mark II with f = 20 mm and f = 200 mm prime focal length lens, and an f = 400 mm focal length lens. Photographs were collected from the opposite valley slope, near the town of Casso (which was only very slightly impacted by the landslide wave in 1963) from a distance ranging between 800 m and 2000 m. The analysis of the 3D models evidenced a progressive variation of the slope dip and dip direction from east (ranging between 46°/ 002° and 36°/ 353°) to west (34°/ 031°), suggesting that the basal rupture surface is characterized by a "bowl" shape due to a synform parallel to the Massalezza Creek. Discontinuity mapping was performed using the 3D models, and a total of seven discontinuity sets, including the bedding, were identified [79] (Figure 7b–e). A rotation of the discontinuity sets was noted, from the east to the west, which is also related to the Massalezza Creek fold [79].

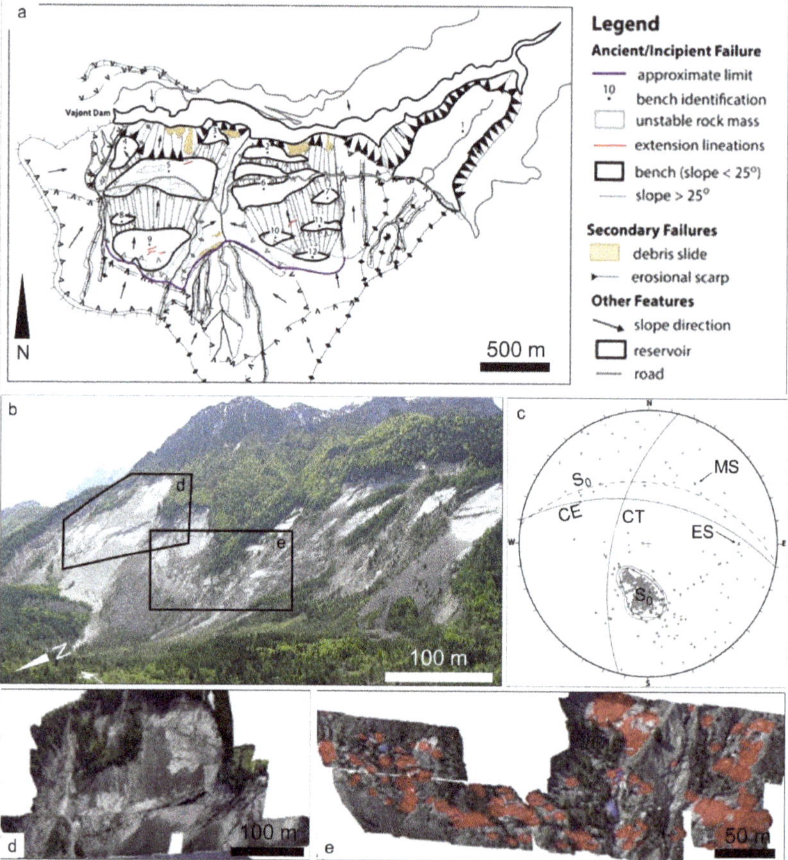

Figure 7. Overview of the remote sensing analysis undertaken at the Vajont Slide site. **a**: geomorphological analysis of the pre-1963 slope, based on historic aerial photo-interpretation; **b**: oblique image of the failed slope and slide deposit. The boxes shows outline the location of the 3D TDP models shown in c–e; **c**: lower hemisphere stereonet plot showing the discontinuity planes mapped across the rupture surface; **d**: 3D TDP model of part of the eastern basal rupture surface, constructed using a f = 200 mm focal lens length; **e**: 3D TDP model of the Massalezza Creek area, constructed using a f = 400 mm focal lens length (**a,c–e** are modified from [79], by permission).

A large-scale morphological analysis of the sliding surface was undertaken using the long-range TDP 3D models, and variations in roughness, concavity, and convexity throughout the sliding surface were observed. Based on this analysis, a semi-quantitative classification of the daylighting sliding surface of the Vajont Slide was proposed [74].

Finally, a qualitative analysis of the evolution with time of the rupture surface was also performed. The release of material from the unstable slope is documented using oblique photographs consistently collected from the opposite slope of the Vajont gorge by various authors in 1964, 1985, 1993, 2002, and 2011 [74].

4. Discussion

The remote sensing analyses conducted at the selected sites and reviewed in this paper allowed better understanding of the geological and geotechnical characteristics at each site. Compared to traditional field-based investigations alone, remotely-sensed geomechanical data could be collected from inaccessible parts of the slope, allowing a more comprehensive study to be undertaken.

At the Hope Slide, traditional field-based rock mass characterization was performed only at the base of the lateral scarp, along the daylighting part of the failure surface, and in the area behind the headscarp [45,47]. Whereas fieldwork results, together with aerial photographs interpretation allowed the recognition of the strong structural control on the 1965 failure, the progressive change in discontinuity set orientation across the slope only became clear after the long-range TDP survey. Future analyses may focus on investigating the role of this structural configuration on the stability and pre-failure deformation of the slope. The use of SfM to reconstruct the pre-failure topography with historical aerial photographs allowed a robust volume estimation to be undertaken, and a 3D geometry for numerical modelling investigations was also obtained. IRT allowed discontinuity seepage to be mapped from long range. Repeated surveys may be undertaken to investigate seasonal groundwater fluctuation and seepage distribution throughout the slope. A similar analysis undertaken using field-based techniques would require strenuous hikes on slippery surfaces, especially after rainfall events and during snowmelt.

Remote sensing analysis conducted at the Block 731 focused on the characterization of the slope from a geological and geomechanical perspective. At this site, only the lowermost zone of the slope is accessible. Remote sensing techniques allowed discontinuity orientation data to be collected from areas, which would otherwise only be accessible using ropes. The short survey range allowed for a high-density 3D point cloud to be collected, enhancing the visibility of features at the centimeter scale. HSI was noted to be effective for detailed lithological mapping at this site in view of (a) the short range of the survey, which allowed for high resolution images and (b) the accessibility of the slope, which allowed calibrated targets to be the positioned on the investigated surface. Future analysis will include (a) the identification of the end-members and (b) the comparison of their spectral signature with a spectral library database to reliably identify the mineralogical composition of the rock.

At the Palliser rockslide site, two orthogonal scanline surveys were conducted at the base of the slope. Such a configuration was critical to limit orientation bias [61]. The ability to perform scanline mapping in the field is strictly dependent on the availability and orientation of accessible outcrops. TDP 3D models significantly increased the number of locations that could be used for discontinuity mapping and allowed window mapping to be undertaken. The analysis of TDP 3D models at the Palliser rockslide allowed a detailed characterization of the discontinuities in inaccessible areas to be performed, including the analysis of bedding plane waviness. TLS surveys conducted in 2017 also provided high resolution point clouds. Compared to TDP models and high resolution photographs, the TLS datasets also include reflectance data. Reflectance represents the ratio of the reflected pulse amplitude to the amplitude value of a diffusely reflecting white target, and is independent of distance to the slope [80]. Its value is affected by various factors, such as weathering and moisture, and, therefore, it allows preliminarily analysis and mapping of surface alteration and discontinuity seepage. Future work at the Palliser rockslide will comprise large-scale structural analyses, including the characterization of the Serrail Creek Syncline, and further small-scale, window discontinuity mapping and rock mass characterization of the rock slope. Additionally, using an integrated TLS-high resolution photograph dataset, a 3D analysis will be undertaken to investigate the areal percentage of failed rock bridge, and to examine its variations throughout the rupture surface.

Geological investigations conducted at the Frank Slide site included field-based analyses (discontinuity mapping, rock mass quality estimation, intact rock strength estimation) performed behind the 1903 headscarp. The rupture surface of the slide is inaccessible due to slope steepness. The long-range TDP survey allowed 3D models of the slope to be constructed and used to perform rock mass characterization. TLS investigations were also undertaken. Due to the limited range of the Optech ILRIS3D laser scanner, however, the survey stations were located in the area of North Peak, providing a better line of sight for the analysis of the South Peak area. The use of a longer-range instrument, such as the VZ-4000, allowed a TLS survey to be performed from various locations within the slide deposit, from a distance of over 2 km. An IRT survey was also performed from the Frank Slide Interpretive Center. It was noted that the survey distance affects the resolution of the dataset but does not prevent the collection of thermal data. Therefore, IRT potentially allows sufficiently extended areas (possibly, three or four times the ground pixel size) characterized by discontinuity seepage to be identified and mapped at very long range. Likewise, a HSI survey allowed a low resolution dataset to be collected. Although lithological variations were observed across the slope, quantitative analyses could not be undertaken due to (a) the large ground pixel size of the dataset, which causes the spectral response of the rock to be averaged over a large area, and (b) to the lack of spectrally calibrated targets on the rock face.

The field-based geological investigation at the Vajont Slide predominantly involved the detailed, geomorphological characterization of the slide deposit. Accessible locations for field mapping along the rupture surface were, instead, limited, and located on steep, scree-covered slopes. Mapping of small-scale structures (folds, faults, jointing) could be undertaken at these locations. Additionally, the analysis of centimeter-scale clayey layers that contributed to the 1963 event was possible only by direct observation of the outcrops. Conversely, comprehensive discontinuity mapping of inaccessible slopes, including the sliding surface and the sub-vertical, eastern, lateral release surface, could only be performed using long-range TDP surveys; 3D TDP models were also employed to perform advanced topographic analyses in ArcGIS and Polyworks, highlighting the bowl-shaped morphology of the rupture surface.

5. Summary: Experience Gained and Lessons Learned

Throughout the past 15 years, many rock slopes and rockslide sites have been investigated in Canada and abroad using field methods and multiple remote sensing techniques. This extensive experience has allowed us to gain knowledge and understanding of the advantages, limitations, and potential uses of the various techniques both at short and long range.

5.1. ALS, UAV-SfM, Aerial Imagery

Aerial datasets allow geoscientists to perform preliminary, large-scale structural, and geomorphic characterization of the investigated area, as well as preliminary, low-resolution change detection analyses. A great advantage of ALS, compared to UAV-SfM and aerial imagery, is the possibility of providing BE datasets in which vegetation and human structures are digitally removed from the dataset. This allows for structural lineaments and geomorphic features to be identified and mapped even on vegetated slopes. Conversely, structural investigations using UAV-SfM and aerial imagery are reliable almost exclusively on non-vegetated surfaces, or for the mapping of large geomorphic features. UAV-SfM, however, can provide high-quality datasets at a relatively low cost. The dataset resolution will depend on the flight height and the specifications of the camera. Aerial imagery can also be employed for geomorphic and structural interpretation, including 3D model reconstruction using an SfM approach (depending on the amount and overlap of the photographs), with the additional advantage that historic photographic imagery can be retrieved from public authorities and institutions. In general, aerial techniques are poorly suited for the investigation of vertical or subvertical rock slopes, due to the unfavorable line of sight.

5.2. TDP, SfM, and TLS

The collection of TDP and TLS datasets provides the means for extensive discontinuity data collection, including orientation, persistence, and spacing, at both long- and short-range, as well as performing geomorphologic characterization and change detection analyses. While both techniques provide a 3D point cloud or mesh of the slope, there are key advantages and limitations, depending on the characteristic of the site and the environmental conditions. Sampling bias associated with TDP and TLS discontinuity characterization are discussed in [62].

The accessibility and visibility of the slope is the first aspect that should be considered in order to choose the most appropriate remote sensing method. Our experience shows that TLS provides highly reliable data with the minimum effort. A single scanning station is generally sufficient to provide a 3D dataset adequate for discontinuity mapping and rock mass characterization. Nevertheless, scanning from multiple locations is preferable to avoid occlusions due to vegetation or changes in rock slope surface orientation. Conversely, TDP and SfM always require a minimum of two camera stations, and the resulting 3D model will exclusively comprise surfaces that are visible from both stations. Therefore, the use of more than two camera stations is recommended. This precaution, although beneficial in preventing occlusions due to variable orientation of the rock slope, can sometimes be impractical, due to difficult terrain conditions, limited accessibility, and limited visibility. The use of oblique, aerial-based techniques (e.g., UAV-SfM) may be considered to overcome this limitation.

TLS surveys were found to be effective in collecting 3D data with poor visibility conditions, due to fog, smoke, and/or rain. Although an increased number of unreliable echoes (e.g., mid-air points in the point cloud) are generally obtained with poor visibility conditions, it was noted that cleaning the point cloud with filters typically available in the processing software (e.g., RiScan Pro) is a relatively easy task. Conversely, adequate visibility and weather conditions are critical for a successful TDP and SfM survey. Haze, fog, and smog can severely limit the visibility, especially at a long range, potentially preventing the collection of a reliable dataset. Conversely, completely clear, sunny conditions can cause the presence of shadows along the rock slope surface, which may result, both at long- and short-range, in areas of no-data or artifacts within the 3D model. We suggest that days with high, even clouds represent the best conditions for undertaking TDP and SfM surveys. It is noted that collecting RAW images (i.e., unprocessed, uncompressed files) may provide a means to considerably enhance the quality of the dataset, by (a) brightening the surfaces in shadow and (b) enhancing contrast and color intensity to improve the sharpness of photographs collected in hazy, foggy, or smoky conditions.

The distance from the slope can be critical in determining the technique to be used. Generally, high-resolution digital reflex cameras can provide longer survey ranges, depending on the visibility conditions. However, the ground pixel size (function of sensor size, focal length, and MegaPixel count) will increase linearly with the distance of observation, causing in turn a decrease in the detail of the photograph. In contrast, the maximum range of a TLS survey is a function of the instrument specifications, and it can further decrease if the rock slope comprises low reflectance material (e.g., seepage, dark-coloured rocks). Additionally, the radius of the emitted laser beam increases with the distance as a function of the laser wavelength.. The beam footprint is the surface area of rock slope covered by the laser beam and represents the surface within which the position of the point in the cloud is averaged. This phenomenon, referred to as "beam divergence", is exacerbated by atmospheric scattering, and causes sharp edges on the rock slope to look "smoothed" in dense point clouds collected from very long range. It is stressed that very dense point clouds are not necessarily characterized by higher detail; on the contrary, they may be unnecessarily large (in terms of file size) and provide the surveyor with a false sense of confidence.

5.3. High-resolution Photography

High-resolution photography provides a means to investigate small details within a rock slope from long range. High-resolution imagery may be used for the construction of 3D TPD and SfM models, if imagery is collected from multiple stations. However, the construction of 3D models using very high-resolution imagery (e.g., 50 MegaPixel) may require long runtime and a high computational effort, due to the large file size of this type of imagery. For instance, a single high-quality jpeg image collected using the Canon EOS 5Ds-R may be up to 30 MB, whereas a RAW file may take up to 100 MB. Limitations of high-resolution photography are similar to those of TDP and SfM surveys, with regard to visibility, weather conditions, and presence of shadows along the investigated slope.

5.4. IRT and HSI

Both IRT and HSI can characterize the emitted and reflected infrared radiation of rock slopes,respectively, and they share similar advantages and limitations. IRT and HSI are generally characterized by a relatively low resolution, due to the limited number of pixels within the sensors. This results in large ground pixel size values at long distance, which prevent the observation of small details. However, it has been shown that long-range analyses are still able to provide important information on the large-scale spatial distribution of seepage (see IRT dataset of Hope Slide) and lithological variation (as in the HSI dataset of Frank Slide) within a rock slope. In contrast, the resolution of these techniques allows the collection of high-quality datasets at short range, as observed at the Block 731 site.

IRT surveys are generally easier and faster, compared to HSI. Thermal cameras provide 2D datasets that can be observed in real time, and instantaneously collected. Conversely, ground-based, HSI systems used for geological application (such as the Specim SWIR3) generally comprise push-broom sensors, which consist of a single column of pixels. Therefore, in order to collect an hyperspectral cube [81], the scanner needs to physically move using a rotor, increasing the time required for the survey.

The analysis and interpretation of IRT datasets is straightforward, and no correction is generally required to identify relative temperature differences. The thermal image is exported as a raster image in which each pixel is color-coded based on the temperature of the slope. More advanced investigations, however, can be undertaken by analyzing the temperature variation with time during the cooling cycle of the rock [21]. The interpretation of HSI datasets is more complex and requires familiarity with the processing software and the characteristics of hyperspectral datasets. The output of an HSI survey is a 3D cube that contains the irradiation data of the investigated object. Such a dataset requires a substantial amount of filtering and calibration before any quantitative analysis can be undertaken. An important part of this procedure is the correction of the dataset, using an empirical-line approach [41], which requires one (or, preferably, two) spectrally calibrated targets that should be positioned next to the investigated slope. While this does not present difficulties at short range, at long range multiple issues may arise, including (a) potential difficulties in accessing the investigated slope, (b) the need for larger spectrally calibrated targets, due to the increased ground pixel size, and (c) the need to consider the impact of atmospheric effects if a significant variation in distance exists across the scene.

5.5. Potential Applications of Remote Sensing Techniques

Remote sensing techniques can be instrumental in providing reliable datasets for rock slopes that are inaccessible due to steepness, ongoing instability, and safety concerns in general. The workflow in Figure 8 summarizes the potential applications of each of the described remote sensing techniques in the characterization of rock slopes, emphasizing the scale (i.e., range) of the analysis.

The data collected using both traditional field and remote sensing techniques may be used to perform preliminary and advanced numerical modelling investigation, which can provide a better understanding of the potential failure mechanisms that affect or affected the slope [12].

The interpretation of the collected data and numerical modelling results can be enhanced by using advanced geovisualization methods, such as mixed reality (MR) and virtual reality (VR), which allow for the observation of 3D datasets and numerical simulations in an immersive virtual environment [82].

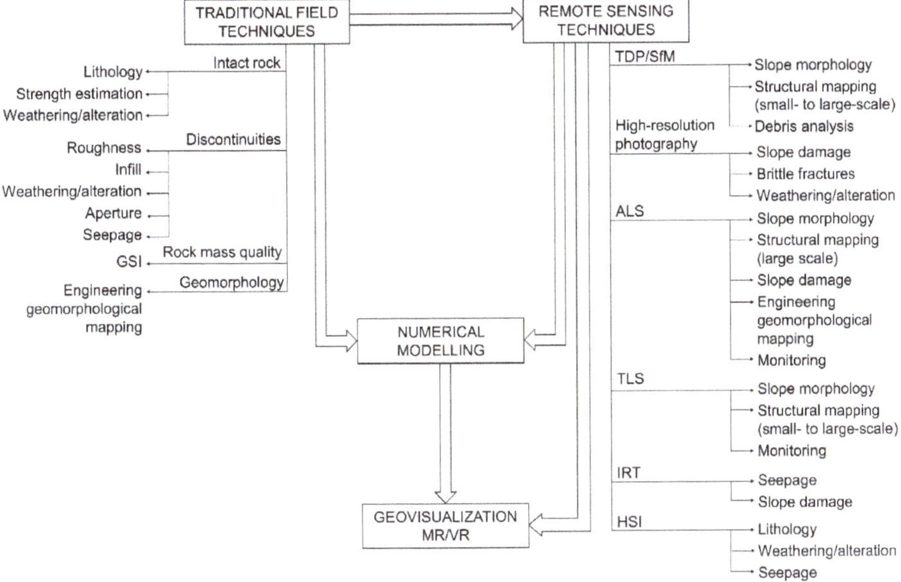

Figure 8. Workflow describing an approach for the comprehensive analysis of rock slopes.

6. Conclusions

The comprehensive characterization of rock slopes is critical to the identification of the failure mechanisms and factors that control their stability. Steepness of the slopes, rock fall activity, difficult terrain, and other safety concerns are some of the factors that may prevent collection of geological, structural, geomorphological, and hydrogeological data using traditional field methods. Therefore, remote sensing surveys often represents a critical task in the characterization and stability analysis of high rock slopes.

Photogrammetric (TDP and SfM) and laser scanner techniques (TLS and ALS) are routinely employed for the characterization of rock slopes and rock masses at various scales. However, the analysis of 3D datasets predominantly focuses on discontinuity mapping, including orientation, persistence, and spacing. Our experience suggests that the integrated use of multiple remote sensing techniques can improve our understanding of the geological and geomorphological processes and factors that govern the stability of rock slopes. Research conducted within the Engineering Geology and Resource Geotechnics Research Group at Simon Fraser University has demonstrated potential applications of various techniques in the investigation of rock slopes and landslides. In this paper, we have provided a summary of some of the sites investigated throughout the past 15 years and highlighted advantages, limitations, and potential application of the various remote sensing techniques employed. Aerial datasets, including ALS, UAV-SfM, and aerial imagery, can be employed to perform large scale, structural and geomorphological characterization of the investigated area. Ground-based methods, such as TDP, SfM, TLS, and high-resolution photography are instrumental in collecting structural and geomorphological data at various ranges, including slope and brittle damage analysis both in 2D and 3D. IRT can be effectively used to investigate and map groundwater seepage. More recently, ground-based HSI has been introduced for the lithological characterization of rock slopes, and potential applications include surface weathering and alteration investigation, shear zone characterization, and rock fall scar analyses.

Despite the successful application of these techniques in the characterization of rock slopes and landslides, we emphasize that traditional field methods are still required to collect extremely important geological and geotechnical data, including intact rock strength, discontinuity infill characteristics, and joint conditions. Therefore, remote sensing techniques should not be considered a substitute for traditional geotechnical field techniques, rather as complementary tools that can greatly enhance our understanding of the processes and mechanisms governing the stability of rock slopes and landslides.

Author Contributions: Conceptualization, Doug Stead; Writing-Original Draft Preparation, Doug Stead and Davide Donati; Writing-Review & Technical Editing, Andrea Wolter and Matthieu Sturzenegger; Supervision, Doug Stead; Funding Acquisition, Doug Stead.

Funding: The authors would like to acknowledge financial support provided through a NSERC Discovery Grant (ID: RGPIN 05817) and FRBC Endowment funds provided to Doug Stead. A KEY Big Data graduate scholarship was provided to Davide Donati.

Acknowledgments: The authors acknowledge the contribution of three anonymous reviewers that improved the quality of the paper. The authors would like to acknowledge Parks Canada for allowing remote sensing surveys to be undertaken within Banff, Jasper, and Glacier National Parks (permit number: JNP-2017-24719).

Conflicts of Interest: The authors declare no conflict of interest.

References

1. Petley, D. Global patterns of loss of life from landslides. *Geology* **2012**, *40*, 927–930. [CrossRef]
2. Seneviratnen, S.; Nicholls, N.; Easterling, D.; Goodess, C.M.; Kannae, S.; Kossin, J.; Luo, Y.W.; Marengo, J.A.; Mcinnes, K.L.; Rahimi, M.; et al. Changes in climate extremes and their impacts on the natural physical environment: An overview of the IPCC SREX report. In *Managing the Risks of Extreme Events and Disasters to Advance Climate Change Adaptation*; Cambridge University Press: Cambridge, UK; New York, NY, USA, 2012; pp. 109–230.
3. Gariano, S.; Guzzetti, F. Landslides in a changing climate. *Earth-Sci. Rev.* **2016**, *162*, 227–252. [CrossRef]
4. Agliardi, F.; Crosta, G.; Zanchi, A. Structural constraints on deep-seated slope deformation kinematics. *Eng. Geol.* **2001**, *59*, 83–102. [CrossRef]
5. Brideau, M.-A.; Yan, M.; Stead, D. The role of tectonic damage and brittle rock fracture in the development of large rock slope failures. *Geomorphology* **2009**, *103*, 30–49. [CrossRef]
6. Havaej, M.; Stead, D.; Mayer, J.; Wolter, A. Modelling the relation between failure kinematics and slope damage in high rock slopes using a lattice scheme approach. In Proceedings of the 48th US Rock Mechanics/Geomechanics Symposium, Minneapolis, MN, USA, 25–28 June 2014; ARMA: New York, NY, USA, 2014.
7. Spreafico, M.C.; Cervi, F.; Francioni, M.; Stead, D.; Borgatti, L. An investigation into the development of toppling at the edge of fractured rock plateaux using a numerical modelling approach. *Geomorphology* **2017**, *288*, 83–98. [CrossRef]
8. Gischig, V.S.; Eberhardt, E.; Moore, J.R.; Hungr, O. On the seismic response of deep-seated rock slope instabilities—Insights from numerical modeling. *Eng. Geol.* **2015**, *193*, 1–18. [CrossRef]
9. Preisig, G.; Eberhardt, E.; Smithyman, M.; Preh, A.; Bonzanigo, L. Hydromechanical Rock Mass Fatigue in Deep-Seated Landslides Accompanying Seasonal Variations in Pore Pressures. *Rock Mech. Rock Eng.* **2016**, *49*, 2333–2351. [CrossRef]
10. Clayton, A.; Stead, D.; Kinakin, D.; Wolter, A. Engineering geomorphological interpretation of the Mitchell Creek Landslide, British Columbia, Canada. *Landslides* **2017**, *14*, 1655–1675. [CrossRef]
11. Leith, K.J. Stress Development and Geomechanical Controls on the Geomorphic Evolution of Alpine Valleys. Ph.D. Thesis, ETH Zurich, Zürich, Switzerland, 2012.
12. Donati, D.; Stead, D.; Brideau, M.-A.; Ghirotti, M. A remote sensing approach for the derivation of numerical modelling input data: Insights from the Hope Slide, Canada. In Proceedings of the 'Rock Mechanics for Africa', Proceedings of the ISRM International Symposium AfriRock Conference, Capetown, South Africa, 3–5 October 2017.

13. Romeo, S.; Di Matteo, L.; Kieffer, D.; Tosi, G.; Stoppini, S.; Radcioni, F. The use of gigapixel photogrammetry for the understanding of landslide processes in alpine terrain. *Geosciences* **2019**, *9*, 99. [CrossRef]

14. Francioni, M.; Salvini, R.; Stead, D.; Coggan, J. Improvements in the integration of remote sensing and rock slope modelling. *Nat. Hazards* **2018**, *90*, 975–1004. [CrossRef]

15. Francioni, M.; Salvini, R.; Stead, D.; Giovannini, R.; Riccucci, S.; Vanneschi, C.; Gullì, D. An integrated remote sensing-GIS approach for the analysis of an open pit in the Carrara marble district, Italy: Slope stability assessment through kinematic and numerical methods. *Comput. Geotech.* **2015**, *67*, 46–63. [CrossRef]

16. Francioni, M.; Coggan, J.; Eyre, M.; Stead, D. A combined field/remote sensing approach for characterizing landslide risk in coastal areas. *Int. J. Appl. Earth Obs. Geoinf.* **2018**, *67*, 79–95. [CrossRef]

17. Abellán, A.; Oppikofer, T.; Jaboyedoff, M.; Rosser, N.J.; Lim, M.; Lato, M.J. Terrestrial laser scanning of rock slope instabilities. *Earth Surf. Process. Landf.* **2014**, *39*, 80–97. [CrossRef]

18. Kromer, R.A.; Hutchinson, D.J.; Lato, M.J.; Gauthier, D.; Edwards, T. Identifying rock slope failure precursors using LiDAR for transportation corridor hazard management. *Eng. Geol.* **2015**, *195*, 93–103. [CrossRef]

19. Vivas, J.B. Groundwater Characterization and Modelling in Natural and Open Pit Rock Slopes. Master's Thesis, Simon Fraser University, Burnaby, BC, Canada, 2014.

20. Kurz, T.H.; Buckley, S.J.; Howell, J.A. Close range hyperspectral imaging integrated with terrestrial lidar scanning applied to rock characterisation at centimetre scale. *ISPRS Int. Arch. Photogramm. Remote Sens. Spat. Inf. Sci.* **2012**, *XXXIX-B5*, 417–422. [CrossRef]

21. Teza, G.; Marcato, G.; Castelli, E.; Galgaro, A. IRTROCK: A MATLAB toolbox for contactless recognition of surface and shallow weakness of a rock cliff by infrared thermography. *Comput. Geotech.* **2012**, *45*, 109–118. [CrossRef]

22. Buckley, S.J.; Kurz, T.H.; Howell, J.A.; Schneider, D. Terrestrial LiDAR and hyperspectral data fusion products for geological outcrop analysis. *Comput. Geosci.* **2013**, *54*, 249–258. [CrossRef]

23. Hoek, E.; Kaiser, P.K.; Bawden, W.F. *Support of Underground Excavations in Hard Rock*; CRC Press: Boca Raton, FL, USA, 1995.

24. ISRM. Suggested methods for the quantitative description of discontinuities in rock masses. *Int. J. Rock Mech. Min. Sci. Abstr.* **1978**, *15*, 319–368. [CrossRef]

25. Barton, N. The shear strength of rock and rock joints. *Int. J. Rock Mech. Min. Sci. Abstr.* **1976**, *13*, 255–279. [CrossRef]

26. Hoek, E.; Brown, E.T. The Hoek–Brown failure criterion and GSI—2018 edition. *J. Rock Mech. Geotech. Eng.* **2019**, *11*, 445–463. [CrossRef]

27. Doornkamp, J.C.; Brunsden, D.; Jones, D.K.C.; Cooke, R.U.; Bush, P.R. Rapid geomorphological assessments for engineering. *Q. J. Eng. Geol. Hydrogeol.* **2007**, *12*, 189–204. [CrossRef]

28. Griffiths, J.S.; Stokes, M.; Stead, D.; Giles, D. Landscape evolution and engineering geology: Results from IAEG Commission 22. *Bull. Eng. Geol. Environ.* **2012**, *71*, 605–636. [CrossRef]

29. Birch, J.S. Using 3DM Analyst mine mapping suite for rock face characterisation. Available online: http://www.adamtech.com.au/3dm/GoldenRocks.pdf (accessed on 20 April 2019).

30. Westoby, M.J.; Brasington, J.; Glasser, N.F.; Hambrey, M.J.; Reynolds, J.M. 'Structure-from-Motion' photogrammetry: A low-cost, effective tool for geoscience applications. *Geomorphology* **2012**, *179*, 300–314. [CrossRef]

31. AdamTechnology. 3DM Analyst Mine Mapping Suite 2.5—User's manual. Unpublished work, 2015.

32. Agisoft LLC. Photoscan 1.4, 2017. Available online: https://www.agisoft.com (accessed on 1 May 2019).

33. Oppikofer, T.; Jaboyedoff, M.; Blikra, L.; Derron, M.-H.; Metzger, R. Characterization and monitoring of the Åknes rockslide using terrestrial laser scanning. *Nat. Hazards Earth Syst. Sci.* **2009**, *9*, 1003–1019. [CrossRef]

34. Riegl LMS GmbH. RiSCAN Pro version 2.6, 2018. Available online: http://www.riegl.com/ (accessed on 20 April 2019).

35. CloudCompare (version 2.10) [GPL software]. 2018. Available online: http://www.cloudcompare.org/ (accessed on 1 May 2019).

36. InnovMetric. PolyWorks 2018 IR10.1. Available online: https://www.innovmetric.com/en (accessed on 20 May 2019).

37. Baroň, I.; Bečkovský, D.; Míča, L. Application of infrared thermography for mapping open fractures in deep-seated rockslides and unstable cliffs. *Landslides* **2014**, *11*, 15–27. [CrossRef]

38. FLIR Systems Inc. ResearchIR 4, 2015. Available online: https://www.flir.com/products/researchir/ (accessed on 20 May 2019).

39. Hunt, G.R. Electromagnetic radiation; the communication link in remote sensing. In *Remote Sensing in Geology*; Siegal, B.S., Gillespie, A.R., Eds.; John Wiley & Sons: Hoboken, NJ, USA, 1980; pp. 5–45.

40. Van Der Meer, F.D.; Van Der Werff, H.M.A.; Van Ruitenbeek, F.J.A.; Hecker, C.A.; Bakker, W.H.; Noomen, M.F.; Van Der Meijde, M.; Carranza, E.J.M.; De Smeth, J.B.; Woldai, T. Multi- and hyperspectral geologic remote sensing: A review. *Int. J. Appl. Earth Obs. Geoinf.* **2012**, *14*, 112–128. [CrossRef]

41. Kurz, T.H.; Buckley, S.J.; Howell, J.A. Close-range hyperspectral imaging for geological field studies: Workflow and methods. *Int. J. Remote Sens.* **2013**, *34*, 1798–1822. [CrossRef]

42. Harris Geospatial Solutions. ENVI 5.5, 2018. Available online: https://www.harrisgeospatial.com/Software-Technology/ENVI (accessed on 20 June 2019).

43. Mathews, W.H.; McTaggart, K.C. The Hope Landslide, British Columbia. *Proc. Geol. Assoc. Can.* **1969**, *20*, 65–75.

44. Weichert, D.; Horner, R.B.; Evans, S.G. Seismic signatures of landslides: The 1990 Brenda mine collapse and the 1965 Hope rockslides. *Bull. Seismol. Soc. Am.* **1994**, *84*, 1523–1532.

45. Von Sacken, R. New Data and Re-Evaluation of the 1965 Hope Slide, British Columbia. Master's Thesis, University of British Columbia, Vancouver, BC, Canada, 1991.

46. Donati, D. The Characterization of Slope Damage Using an Integrated Remote Sensing-Numerical Modelling Approach. Ph.D. Thesis, Simon Fraser University, Burnaby, BC, Canada. In preparation.

47. Brideau, M.A.; Stead, D.; Kinakin, D.; Fecova, K. Influence of tectonic structures on the Hope Slide, British Columbia, Canada. *Eng. Geol.* **2005**, *80*, 242–259. [CrossRef]

48. Donati, D. A Structural Investigation of the Hope Slide, BC, Using Terrestrial Photogrammetry and Rock Mass Characterization. Master's Thesis, University of Bologna, Bologna, Italy, 2012.

49. ESRI. ArcGIS 10.5, 2017. Available online: https://desktop.arcgis.com/en/ (accessed on 20 June 2019).

50. Corkum, A.G.; Martin, C.D. Analysis of a rock slide stabilized with a toe-berm: A case study in British Columbia, Canada. *Int. J. Rock Mech. Min. Sci.* **2004**, *41*, 1109–1121. [CrossRef]

51. Imrie, A.S.; Moore, D.P. The use of rock engineering to overcome adverse geology at Revelstoke Dam. *Surf. Undergr. Proj. Case* **1993**, *5*, 701–725. [CrossRef]

52. Stewart, T.W.G.; Moore, D.P. Displacement behavior of the Checkerboard Creek rock slope. In *Terrain Stability and Forest Management in the Interior of British Columbia*; Jordan, P., Orban, J., Eds.; British Columbia Forest Science Program: British Columbia, BC, Canada, 2001; pp. 178–190.

53. Read, P.B.; Brown, R.L. Columbia River fault zone: Southeastern margin of the Shuswap and Monashee complexes, southern British Columbia. *Can. J. Earth Sci.* **1981**, *18*, 1127–1145. [CrossRef]

54. Martin, C.D.; Tannant, D.D.; Lan, H. Comparison of terrestrial-based, high resolution, LiDAR and digital photogrammetry surveys of a rock slope. In *Rock Mechanics: Meeting Society's Challenges and Demands*; CRC Press: Boca Raton, FL, USA, 2007; pp. 925–932.

55. Green, A.A.; Berman, M.; Switzer, P.; Graig, M.D. A transformation for ordering multispectral data in term of image quality with implications for noise removal0. *IEEE Trans. Geosci. Remote Sens.* **1988**, *26*, 65–74. [CrossRef]

56. Smith, G.M.; Milton, E.J. The use of the empirical line method to calibrate remotely sensed data to reflectance. *Int. J. Remote Sens.* **1999**, *20*, 2653–2662. [CrossRef]

57. McMechan, M.E. Geology and structure cross-section, Peter Lougheed Provincial Park, Alberta. In *Geological Survey of Canada*; Minister of Supply and Services Canada: Ottawa, ON, Canada, 1989.

58. Jackson, L. Terrain inventory of the Kananaskis Lakes Map Area, Alberta. In *Geological Survey of Canada*; Minister of Supply and Services Canada: Ottawa, ON, Canada, 1987; 40p. [CrossRef]

59. Sturzenegger, M.; Stead, D.; Gosse, J.; Ward, B.; Froese, C.R. Reconstruction of the history of the Palliser Rockslide based on ^{36}Cl terrestrial cosmogenic nuclide dating and debris volume estimations. *Landslides* **2015**, *12*, 1097–1106. [CrossRef]

60. Cruden, D.M.; Eaton, T.M. Reconnaissance of rockslide hazards in Kananaskis Country, Alberta. *Can. Geotech. J.* **1988**, *25*, 411. [CrossRef]

61. Sturzenegger, M.; Stead, D. The Palliser Rockslide, Canadian Rocky Mountains: Characterization and modeling of a stepped failure surface. *Geomorphology* **2012**, *138*, 145–161. [CrossRef]

62. Sturzenegger, M. Multi-Scale Characterization of Rock Mass Discontinuities and Rock Slope Geometry Using Terrestrial Remote Sensing Techniques. Ph.D. Thesis, Simon Fraser University, Burnaby, BC, Canada, 2010.
63. Donati, D.; Stead, D.; Onserl, E. New approaches to characterize brittle fracture and damage in fractured rock masses. In Proceedings of the 10th Asian Rock Mechanics Symposium, Singapore, 29 October–3 November 2018; Paper ARMS10-P-0505.
64. Cruden, D.M.; Martin, C.D. Before the Frank Slide. *Can. Geotech. J.* **2007**, *44*, 765–780. [CrossRef]
65. Humair, F.; Pedrazzini, A.; Epard, J.L.; Froese, C.R.; Jaboyedoff, M. Structural characterization of Turtle Mountain anticline (Alberta, Canada) and impact on rock slope failure. *Tectonophysics* **2013**, *605*, 133–148. [CrossRef]
66. McConnell, R.G.; Brock, R.W. *Report on the Great Landslide at Frank, Alberta, Canada, 1903*; Department of the Interior: Ottawa, ON, Canada, 1904.
67. Benko, B.; Stead, D. The Frank slide: A reexamination of the failure mechanism. *Can. Geotech. J.* **1998**, *35*, 299–311. [CrossRef]
68. Moreno, F.; Froese, C.R. *Turtle Mountain Field Laboratory Monitoring and Research Summary Report, 2005*; EUB/AGS Earth Sciences Report; Alberta Energy and Utilities Board: Calgary, AB, Canada, 2006.
69. Froese, C.R.; Moreno, F.; Jaboyedoff, M.; Cruden, D.M. 25 years of movement monitoring on South Peak, Turtle Mountain: Understanding the hazard. *Can. Geotech. J.* **2009**, *46*, 256–269. [CrossRef]
70. Wood, D.E.; Yusifabayov, J.A.; Chao, D.K.; Shipman, T.C. *Turtle Mountain Field Laboratory, Alberta (NTS 82G): 2016 Data and Activity Summary*; AER/AGS Open File Report; Alberta Energy Regulator: Calgary, AB, Canada, 2018.
71. Brideau, M.-A.; Pedrazzini, A.; Stead, D.; Froese, C.R.; Jaboyedoff, M.; Van Zeyl, D. Three-dimensional slope stability analysis of South Peak, Crowsnest Pass, Alberta, Canada. *Landslides* **2011**, *8*, 139–158. [CrossRef]
72. Sturzenegger, M.; Stead, D. Quantifying discontinuity orientation and persistence on high mountain rock slopes and large landslides using terrestrial remote sensing techniques. *Nat. Hazards Earth Syst. Sci.* **2009**, *9*, 267–287. [CrossRef]
73. Semenza, E.; Ghirotti, M. History of the 1963 Vaiont slide: The importance of geological factors. *Bull. Eng. Geol. Environ.* **2000**, *59*, 87–97. [CrossRef]
74. Hendron, A.J.; Patton, F.D. *The Vaiont Slide, a Geotechnical Analysis Based on Geologic Observations of the Failure Surface*; Technical Report; U.S. Army Corps of Engineers Waterways Experiment Station: Vicksburg, Mississippi, 1985; Volume 1.
75. Costa, V.; Doglioni, C.; Grandesso, P.; Masetti, D.; Pellegrini, G.B.; Tracanella, E. Carta Geologica d'Italia alla scala 1:50.000. Note illustrative del F°063 Belluno, 1992; 74p. (In Italian). Available online: http://www.isprambiente.gov.it/Media/carg/veneto.html (accessed on 20 June 2019).
76. Massironi, M.; Zampieri, D.; Superchi, L.; Bistacchi, A.; Ravagnan, R.; Bergamo, A.; Ghirotti, M.; Genevois, R. Geological structures of the Vajont landslide. *Ital. J. Eng. Geol. Environ.* **2013**, *2013*, 573–582. [CrossRef]
77. Wolter, A.; Stead, D.; Ward, B.C.; Clague, J.J.; Ghirotti, M. Engineering geomorphological characterisation of the Vajont Slide, Italy, and a new interpretation of the chronology and evolution of the landslide. *LandSlides* **2016**, *13*, 1067–1081. [CrossRef]
78. Wolter, A. Characterisation of Large Catastrophic Landslides Using an Integrated Field, Remote Sensing and Numerical Modelling Approach. Ph.D. Thesis, Simon Fraser University, Burnaby, BC, Canada, 2014.
79. Wolter, A.; Stead, D.; Clague, J.J. A morphologic characterisation of the 1963 Vajont Slide, Italy, using long-range terrestrial photogrammetry. *Geomorphology* **2014**, *206*, 147–164. [CrossRef]
80. Riegl. 3D Terrestrial Laser Scanner RIEGL VZ-4000—General Description and Data Interfaces, 2014. Available online: http://www.riegl.com/nc/products/terrestrial-scanning/produktdetail/product/scanner/30/ (accessed on 20 June 2019).
81. Shaw, G.A.; Burke, H.K. Spectral imaging for remote sensing. *Linc. Lab. J.* **2003**, *14*, 3–28.
82. Onsel, E.; Donati, D.; Stead, D.; Chang, O. 2018 Applications of virtual and mixed reality in rock engineering. In Proceedings of the 52nd US Rock Mechanics/Geomechanics Symposium, Seattle, WA, USA, 17–20 June 2018.

 International Journal of
Geo-Information

Article

Application of Remote Sensing Data for Evaluation of Rockfall Potential within a Quarry Slope

Carlo Robiati [1,*], Matt Eyre [1], Claudio Vanneschi [2], Mirko Francioni [3], Adam Venn [4] and John Coggan [1]

1   Camborne School of Mines, University of Exeter, Penryn TR10 9FE, Cornwall, UK
2   CGT Spinoff Srl, Via Vetri Vecchi 34, 52027 S. Giovanni Valdarno, AR, Italy
3   Department of Engineering and Geology, University of Chieti-Pescara, 66100 Chieti, Italy
4   Imerys Minerals Ltd., St Georges Rd, Nanpean PL26 7XR, Cornwall, UK
*   Correspondence: c.robiati@exeter.ac.uk

Received: 12 July 2019; Accepted: 19 August 2019; Published: 22 August 2019

Abstract: In recent years data acquisition from remote sensing has become readily available to the quarry sector. This study demonstrates how such data may be used to evaluate and back analyse rockfall potential of a legacy slope in a blocky rock mass. Use of data obtained from several aerial LiDAR (Light Detection and Ranging) and photogrammetric campaigns taken over a number of years (2011 to date) provides evidence for potential rockfall evolution from a slope within an active quarry operation in Cornwall, UK. Further investigation, through analysis of point cloud data obtained from terrestrial laser scanning, was undertaken to characterise the orientation of discontinuities present within the rock slope. Aerial and terrestrial LiDAR data were subsequently used for kinematic analysis, production of surface topography models and rockfall trajectory analyses using both 2D and 3D numerical simulations. The results of an Unmanned Aerial Vehicle (UAV)-based 3D photogrammetric analysis enabled the reconstruction of high resolution topography, allowing one to not only determine geometrical properties of the slope surface and geo-mechanical characterisation but provide data for validation of numerical simulations. The analysis undertaken shows the effectiveness of the existing rockfall barrier, while demonstrating how photogrammetric data can be used to inform back analyses of the underlying failure mechanism and investigate potential runout.

Keywords: rockfall hazard; slope stability; remote sensing; LiDAR; SfM-MVS; photogrammetry

---

1. Introduction

Rock fall during quarrying activities are among the most critical risks associated with slope instability, especially for high cuts in weathered rock [1]. Legacy slopes, such as the one studied in this investigation, may be particularly prone to rockfall, since they were created prior to the UK Quarry Regulations (1999) [2] and regular maintenance may be difficult to undertake. Rockfall is a slope process involving the detachment of rock fragments and their fall and subsequent bouncing, rolling, sliding, and deposition, where the main responsible factor for the rockfall behaviour is the slope inclination and its irregularities [3–5]. Cruden and Varnes [6] define rockfall as a failure where "little or no shear displacement takes place and the material descends mainly through air in free-fall, leaping, bouncing or rolling. Movements are very rapid and may or may not be preceded by minor movements leading to progressive separation of the rock mass from its source." According to McCauley [7], the main cause for the initiation of rockfalls can be directly related to water, namely rain, snow-melt, springs and seeps, and the associated increased pressure due to water infiltration in pores and discontinuities. The triggers and conditions that instigate rockfall that are not directly related to water are root wedging, excavation activities and earthquakes, and these can account for a significant portion of the observed rockfall failures [8]. Rockfall instability phenomena, in natural and engineered slopes, have been under investigation since 1960s, and the results have been published by a large number of researchers; dealing with the physical basis of the process [9–12], and the hazard and risk associated with it [13–16].

In the last two decades, developments in the area of geo-information, in particular in the production of three-dimensional models, has enhanced the ability to carry out rockfall risk assessments in previously inaccessible locations. In addition, workflows have been defined based on the acquisition of 3D point clouds and the extraction of the information that they contain [1,17–21]. Depending upon the frequency of data acquisition and the rate of change of the slope prior to its failure it is possible to hypothesise the slope failure mode, the potential volume of the eventual failure and, in some cases, provide an accurate estimate of the time of failure [22]. In back analysis, it is possible to determine the position of the source areas, to assess the path of movement or trajectory, to calculate the volume of accumulated debris, and the velocity and energy associated with its descent [23]. Point clouds produced directly with laser scanning (LiDAR), terrestrial (TLS) or aerial (ALS) [4,24–26] or produced by means of applying Structure from Motion Multi-View Stereo (SfM-MVS) algorithms, using platforms as Unmanned Aerial Vehicle (UAV) [27–30] can play a fundamental role in the characterisation of rockmass quality and its features. Indeed, both TLS and UAV are easy and fast to operate, allowing one to acquire data with high geometric and temporal resolutions. The identification and localisation of discontinuities enable detailed spatial modelling of these structures that can be used as input parameters for numerical simulations of rock slope stability.

This study provides a rockfall evaluation for a legacy slope in an active open pit operation through back-analysis of the rockfall using remotely captured data. Legacy slopes often do not comply with the most recent regulations, in terms of acceptable geometry (height and profile) and maintenance (not being subject to regular scaling), so the uncertainty of parameters associated with triggering and predisposing factors, as well as purely geometric characteristics (frequency and distribution of discontinuities) render such analyses challenging. However, the availability of high resolution remotely captured datasets (ALS, TLS and SfM-MVS derived point clouds), obtained from surveying campaigns routinely carried out in active mining operations can be of key importance for defining the geological conditions and structures that promote rockfall events. Through definition of the slope's geo-mechanical characteristics and incorporation of this information in a GIS environment, using 2D and 3D numerical simulations, a rockfall reach probability map showing the location and distribution of modelled rockfall has been produced. The influence of the spatial resolution of the surface topography on the modelled behaviour is also investigated. The simulations were run using an array of topographic models at different resolutions to investigate the effect of grid size on the run-out of rock blocks. A specific test was designed to study the effect of rock blocks' shape in the 3D simulations. Part of this investigation is also to ascertain if the protection measures are adequately designed to mitigate the risk resulting from further potential detachments from the rock face. The study demonstrates the benefits of using probabilistic numerical models, both in two and three dimensions, in order to reduce the uncertainty arising from the difficulties of validating results in inaccessible areas. As far as the Authors are aware, there are no studies showing a direct comparison of rockfall trajectory analysis models in 2D and 3D, and how this comparison can lead to a more reliable assessment of rockfall hazard.

Case Study: The Treviscoe Rockfall

The case study is based on back analysis of a rockfall event that initially occurred in 2011, developed into a full slope scale activity in 2013, including a major event in early 2016 with further ongoing activity. The rockfall is located in a section of a quarry bench in Treviscoe Pit, St.Austell, UK (Figure 1a–c). The site location requires a detailed analysis, given that the internal geotechnical assessment suggests that it poses a significant hazard, as per the UK Quarry Regulations (1999) [2]. Treviscoe Pit is located within the St. Austell granite cupola, in SW England. The pit, from which both kaolinite (china clay) and aggregates are extracted, is one of the oldest in the region, resulting in an excavation 600 by 300 m and approximatively 80 m in depth. The rockfall has undergone several different phases of activity, as reported by mine personnel. In order to reduce the risks associated with rockfall, a rock trap was created at the base of the slope to prevent blocks from reaching the haul road.

As can be observed in Figure 2a,b, which illustrates the conditions at the slope prior (2013) and after (2018), the major collapse event occurred in early 2016, and the outcrop presents itself as a fairly irregular sub-vertical rock face. The material outcropping is a weathered topaz-bearing granite [31,32], which has undergone partial to complete kaolinisation. The rockmass appears to contain near vertical columnar joints, spaced 1 m, resting on basal planes gently dipping towards the free face. The sub-vertical wall is approximately 21 m in height. At the base of the sub-vertical section is a slope with a bench angle of approximately 50° that consists of vegetation and scree. The scree is the result of several rockfall events depositing material towards the bottom of the slope. To protect and mitigate against rock fall a sand bund was erected at the base of the slope, approximately 3 m in height, along the whole section of the haul road.

Figure 1. Location of the rockfall slope (**a**) aerial location of the case study position in the overall pit; (**b**) localised aerial view; (**c**) view from the base of the slope. The photographs were acquired in August 2018. Scale bar in (**c**) is indicative.

Figure 2. Photographic comparison of the rockfall (**a**) October 2013 and (**b**) August 2018. The photos were taken from the opposite side of the pit, circa 300 m away from the target and they show the change in geometry due to the major collapse occurred in early 2016. Scale bar is indicative.

2. Materials and Methods

In order to investigate and evaluate the rockfall events several data collection surveys were undertaken and are summarised below. The obtained datasets were utilised to inform different types of analyses, extracting key geo-mechanical parameters and to constrain numerical modelling simulations, i.e., 2D and 3D rockfall trajectory analysis.

2.1. Close-Range Remote Sensing Survey

To generate a DEM (Digital Elevation Model) and to extract geo mechanical information for the rockfall numerical simulations, two remote sensing techniques were selected to capture the scene at close-ranges: TLS and drone-borne SfM-MVS. Given the restrictions posed by the hazardous conditions on the rock wall, the position on the ground from where the LiDAR unit could be operated were limited. The most reasonable choice to overcome this condition after the initial scans was to opt for a drone-based survey, with the capacity of getting closer observations, hence improving the GSD (Ground Sampling Distance), and eventually allowing to capture some previously shadowed zone of the slope, as the top of the cliff and the inner side of the rock-trap. The TLS survey was carried out with a Leica ScanStation C10™, a time-of-flight laser scanner with a nominal scan resolution, at ranges from 0 to 50 m, of 4.5 mm, with an accuracy on a single measurement position and distance of 6 and 4 mm, respectively. The rock face was scanned in May 2017 with a Leica ScanStation C10™, from four individual scan positions, approximatively 35 m away from the target, to obtain a high resolution point cloud of 27×10^6 points, with an average spacing of 0.025 m (shown in Figure 3). The point cloud was registered using a total of five targets, using the Cyclone™ software released by the same manufacturer as the aforementioned LiDAR unit. The TLS point cloud was oriented to magnetic North with the help of a compass bearing. The root-mean-square error (RMSE) on the z coordinate for the point cloud registration is 0.017 m. A second survey was performed in August 2018, as soon as a UAV was available to the mine operation survey team, and a second high resolution point cloud reconstructed by means of SfM-MVS workflow, using optical images taken from a drone. The dataset for the SfM-MVS scene reconstruction was obtained with a DJI Phantom 4 Pro™, using the built-in 20 MP camera with mechanical shutter, controlled remotely by the drone operator. The drone was flown manually, at an average distance of ~20 m from the rock face, acquiring 303 images and using four fixed GCP (Ground Control Point) present in the observed scene. GCP network was established by the mine survey department, having their coordinates measured using D-GNSS (Differential-Global Navigation Satellite System). The point cloud was processed in Agisoft Photoscan™, resulting in a point cloud of 19×10^6 points, with an average point spacing of 0.029 m (shown in Figure 4). Other than a DEM, as an output from the SfM-MVS workflow, a high resolution RGB orthophoto was generated, having a GSD of 0.0073 m. The orthophoto was generated at the highest possible resolution to enable a detailed mapping of the rockfall debris.

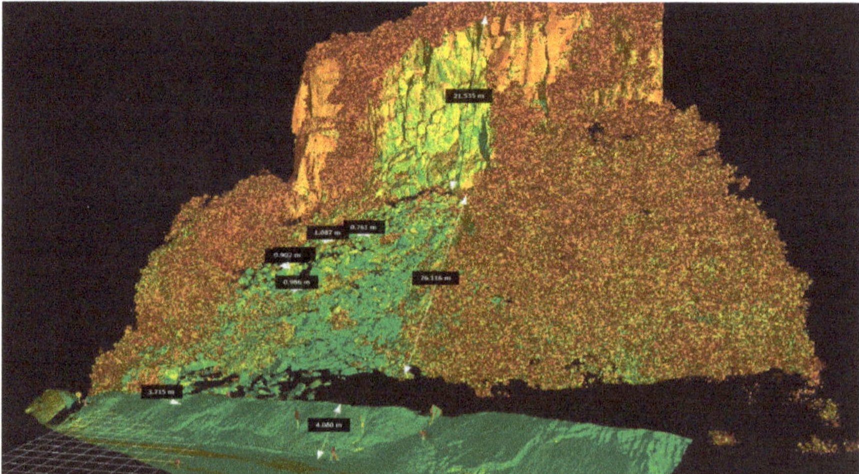

Figure 3. Registered TLS (terrestrial laser scanning) point cloud image, with illustrative sample dimensions highlighting the sub-vertical rock wall and the scree slope. Given the inaccessibility of the inner side of the rockfall trap, the TLS was unable to capture its complete geometry. Some sample measurements were taken and highlighted as to show the indicative vertical drop from the source area to the base of the slope, and the length of the scree slope/transition zone. The same measurements were used to obtain a representative block size on the rock wall.

Figure 4. Coloured Structure from Motion Multi-View Stereo (SfM-MVS) point cloud image, acquired using a Phantom 4 DJI platform equipped with a 20 MP RGB camera and processed in Agisoft Photoscan™. The Unmanned Aerial Vehicle (UAV)-hosted sensor enabled the complete reconstruction of shadowed regions in the scene in the TLS survey.

2.2. Long-Range Remote Sensing Survey

The mining operation, through a contractor, had previously undertaken a series of aerial laser scanning (ALS) derived point cloud, capturing the geometry of the entire pit at 1 m ground sample distance (GSD). The error associated with the z coordinate for the surveys did not exceed 0.070 m for all the campaigns (2011, 2013, 2015, 2016) with a RMSE of 0.060 m that resulted from averaging the RMSE in z of the four campaigns, computed over 52 GCP scattered across the mining operation, i.e., the local reference grid used by the surveying personnel. The multi-temporal, yearly, LiDAR coverage (2011 to 2016) provided the basis to reconstruct the variation in geometry of the pit walls over time. The ALS surveys were performed in the same flights designed to acquire aerial photographs. In this study, aerial photographs served the purpose of changing detection analysis, and to constrain, both temporally and geometrically, the evolution of the rockfall. The orthorectified aerial images (shown in Figure 5) were generated with a GSD of 30 cm and the surveyor's ground truth report indicates a RMSE on the xy plane of 0.15 m. In Figure 5, a series of orthorectified images (from 2011 to 2016), it can be observed how the rockfall, after the initial activity in 2011, occurred through further release events.

The ALS, TLS and SfM-MVS point clouds were rasterised in CloudCompare™, using the rasterize tool, to generate DEM with different grid sizes; using ALS data for GSD above 1 m, TLS data for GSD of 1 m, finally SfM-MVS derived data for topographical models below 1 m GSD. In this way there several DEM, having 5, 2, 1, 0.5 and 0.3 m grid sizes were produced.

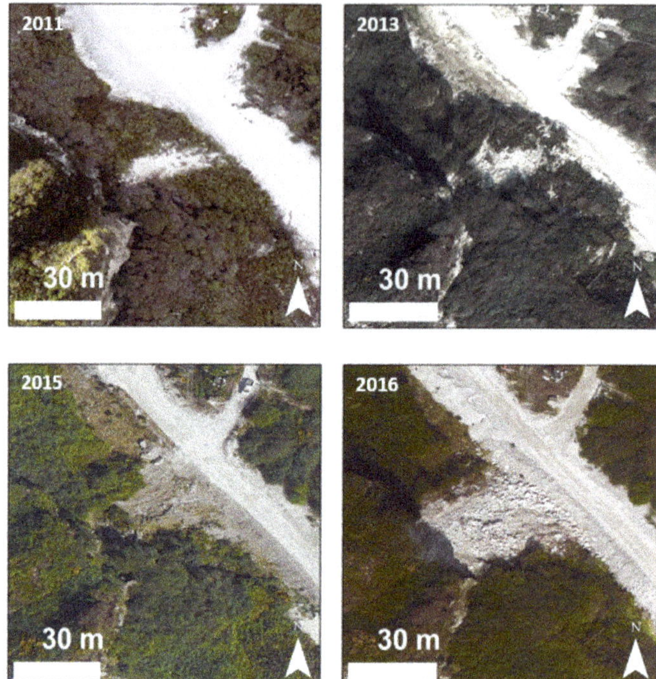

Figure 5. Orthorectified aerial imagery of the rockfall area. The sequence shows the activity of the rockfall, from its initial condition (**top left**), to the development of the rockfall at the whole slope scale in 2013 (**top right**), and the subsequent maintenance operations, the emplacement of the rock trap and the removal of rock fragments from the ditch. The last picture shows the aftermath of the major collapse that led to the actual slope geometry.

2.3. Geo-Mechanical Analysis

In potentially hazardous environments, such as a legacy slope, it is not possible to carry out traditional geo-mechanical surveys due to the unacceptable level of risk the surveyors would be exposed to. Technological advancements in remote mapping platforms have helped to overcome this issue, as reviewed in Tannant [33] and Giordan [29], resulting in a rapidly growing market of reliable, low cost UAV. It has been previously shown that surface topography reconstructed with TLS and SfM-MVS methods can be successfully used to determine the orientation and spacing of rock discontinuities [34]. In order to obtain a statistically robust representation of the discontinuities in the outcropping rockmass, the TLS point cloud was processed and analysed in SplitFX™. This software is designed to extract geo-mechanical information from point clouds by mapping either planar facets or trace planes and subsequently plotting them on a stereoplot. The TLS point cloud data was imported into SplitFX and the discontinuities mapped by manually assigning best fit polygons onto the point cloud surface. The point normal of the polygon can then be represented on an equal area hemispherical stereonet to identify discontinuity sets and perform kinematic analysis for potential failure mechanisms.

The orientation of the rock discontinuities captured within SplitFX was then imported in DIPS™, and a kinematic analysis was performed [8,35]. The analysis of source areas for rockfall includes identification of kinematically admissible unstable blocks and an investigation of the factors influencing the stability of such blocks [36]. The Markland test [35] considers the possible slope failure mechanisms, without considering forces, and has been used to establish some possible rockfall scenarios for further numerical simulation.

2.4. GIS Geospatial Analysis

The geospatial datasets (aerial ortho-photographs, ALS, TLS and SfM-MVS derived DEM) were incorporated in the ESRI ArcGIS™ environment for the purpose of establishing a database of the information related to the rockfall event. The aerial photography coverage from 2011 to 2016 (Figure 5) and the digital photogrammetric survey (both the high resolution orthorectified image and the SfM-MVS derived DEM), enabled the mapping of the activity of the main rockfall scarp, other than the end locations of rock blocks. The aerial imagery time series also served the purpose of recording the effects of maintenance operations, i.e., the creation of the sand embankment protecting the haul road and the removal of rock fragments trapped in the rockfall ditch. The GIS environment provided the platform to analyse and map, in high-resolution, homogeneous areas, namely the (a) source area, the (b) scree slope, the (c) vegetated slope and the (d) rock-trap, shown in Figure 6. GIS evaluation also provided the chance to record the position of end point locations for validation of subsequent rockfall simulations. The spatial analysis toolbox provided means for managing DEM, from which a representative vertical cross section (profile) was taken (shown in Figure 6) for subsequent numerical modelling.

The GIS also formed the basis for pre-processing the rasters used as topography for the 2D (by extracting a vertical profile) and 3D (as an ASCII DEM) numerical simulations through assigning input parameters to the different homogeneous areas mapped, as described in the Section 3.3.

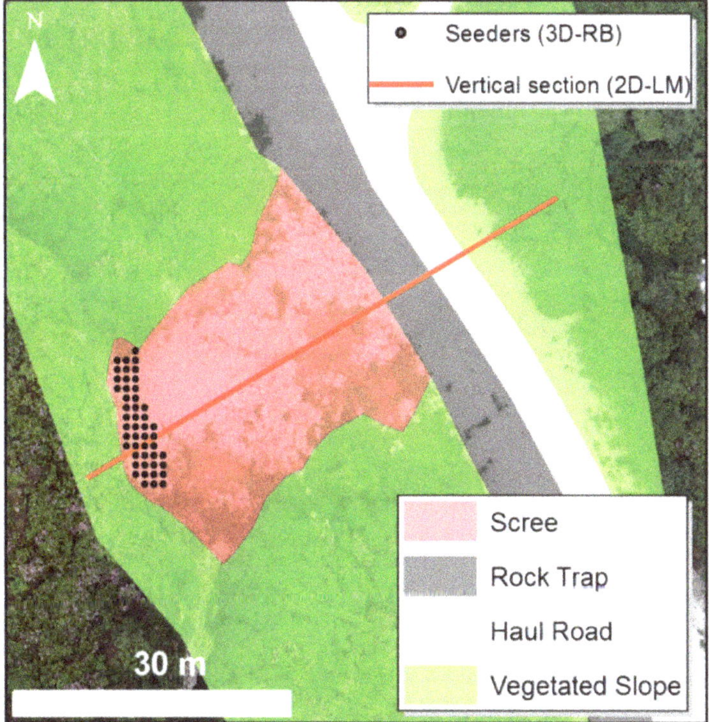

Figure 6. Geo-mechanical zonation used to assign input parameters for the 3D numerical modelling. In red is shown the trace of the representative vertical profile extracted from the Digital Elevation Model (DEM) to obtain the slope geometry for 2D numerical modelling.

2.5. Numerical Modelling

An established method to assess the hazard posed by rockfalls is the use of probabilistic rockfall trajectory analysis [12,37]. A probabilistic approach is adopted to consider both ontic and epistemic uncertainty in rockfall trajectory modelling, i.e., the variability of the information gathered during the surveying phase of the study and the preparation of GIS data layers [36,37]. To produce a realistic simulation of the rockfall behaviour, the models must incorporate a digital representation of the topography, either in the form of a DEM raster or a vertical profile, both of which can be retrieved from remotely sensed data. The topography gradient will govern the general direction that a block will take through its descent. Predefined physical-mechanical characteristics of the digital surface are used to compute the loss of energy for the inelastic rebounds with the ground at each pixel of the DEM. These parameters are called Coefficients of Restitutions (COR), defined as an energy transfer function, which is generally expressed in the form of a ratio between the velocity before and after an impact [10,38]. COR are defined in the normal direction (CORN) and tangential (CORT) to the slope. They are used to account for energy lost due to the inelastic deformation during the collision of a rock with the slope or bench [39]. COR are key parameters for rockfall modelling, and it is necessary to use engineering judgement when selecting appropriate values from literature, especially given the inherent difficulties of defining them empirically through field testing [9,10,40].

A common distinction in how rockfall modelling software treats impact theory is the lumped mass (LM) approach versus the rigid body (RB) approach. The lumped mass approach considers the mass being concentrated in a single point, while the rigid body approach uses a defined geometry to model the rock block. Due to ongoing activities in the pit it was not possible to undertake in-situ field calibration tests for assessing directly the reliability of the models and the effectiveness of the rockfall protection [41,42]. However, validation was obtained from comparison of modelling results with aerial images of the rockfall body and known end locations for rockfall fragments. COR values for the two-dimensional lumped-mass impact model (2DLM) were obtained from literature and compared with a soil cover map of the site.

In this study two software were selected to perform the trajectory analyses: Rocfall™, developed by Rocscience, and Rockyfor3D™, developed by ecorisQ Association. They were selected as they are reliable tools, largely used both in academia and in the industry. In addition, they offer different solutions in terms of statistical assumptions and results typology.

Rocfall™ is a 2D-LM (Lumped Mass) probabilistic, processed-based software for rockfall simulation [43] reproducing the trajectory of rock blocks falling along a 2D slope. The input parameters (i.e., CORN, CORT, static friction, rolling dynamic friction and slope roughness) can be obtained from literature and previous calibrated simulations or example data from the help documentation in the software itself. The software then allows definition of the statistical variability of these input parameters [37]. The point cloud geometries were used as the basis to provide a representative sectional line for simulation. A representative vertical section (shown as a red trace in Figure 6) was generated from both the ALS, TLS and SfM-MVS point clouds and exported into AutoCAD™ to provide the geometry for subsequent 2D rockfall simulations. The vertical section was extracted at different geometrical resolution using the 3D Analyst toolbox in ArcGIS™.

The vertical sections were then traced to form a polyline in AutoCAD™, which was then exported into Rocscience Rocfall™ for analysis. A sensitivity analysis on the slope geometry resolution was undertaken, extracting the topography from ALS, TLS and SfM-MVS derived DEM, at 5, 2, 1, 0.5 and 0.3 m.

Rockyfor3D™ is a three-dimensional rigid-body impact model (3D-RB) (Rigid Body) that calculates trajectories of single, individually falling rocks with discrete geometry (RB), in three dimensions (3D). The model combines physically-based, deterministic algorithms with stochastic approaches, which makes Rockyfor3D a so-called 'probabilistic process-based rockfall trajectory model'. Rockyfor3D can be used for regional, local and slope-scale rockfall simulations [44]. In this software the input parameters are assigned to the digital surface through pre-processing of ASCII GIS data layers (i.e., release cells location, density, shape and dimensions of rock blocks and their statistical variation range and initial vertical velocity). The local slope surface roughness is represented by a parameter defined as maximum obstacle height (MOH), expressed in metres. Typical MOH values, as suggested by Dorren [44], which are encountered by a falling rock are represented by statistical classes, namely rg70%, rg20%, and rg10%. During each rebound calculation, the MOH value in a cell is randomly chosen from the three representative values according to their probabilities of occurrence [44]. A sensitivity analysis on the influence of the DEM resolution was performed for the case study, comparing the results of the rockfall (blocks) end point(s) when using a DEM having a 5, 2, 1, 0.5 and 0.3 m GSD and for different rockfall size scenarios, for 2D-LM and 5, 2 and 1 m for 3D-RB simulations.

3. Results

Following the remote sensing data collection campaigns, the collected data were processed in several software applications to generate the products which were used to classify the slope and evaluate the potential of a rockfall at the site.

3.1. Geo-Mechanical Analysis

Figure 7 illustrates an example of a discontinuity mapping carried out on the point cloud in SplitFX. The mapping identified 348 different entries that were then exported into Rocscience DIPS™ to identify the major discontinuity set orientations characterising the rock mass and to perform kinematic analysis. Figure 8 illustrates the kinematic analysis undertaken in DIPS, where the principal discontinuity sets were identified. Table 1 summarises the major sets orientations represented using Dip and Dip Direction format. The stereographic analysis confirmed the presence of near vertical columnar joints represented by Sets 1 and 2 and basal planes (Sets 3 and 4) dipping out of the face.

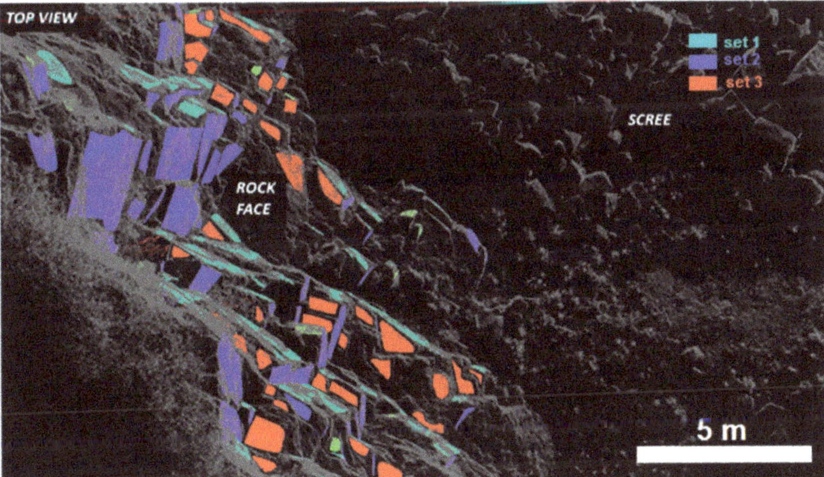

Figure 7. Top view of the rock wall. The coloured facets highlight the discontinuity network identified in Figure 8. The blue and teal patches identify joint Sets 1 and 2, responsible for isolating columnar-like prisms and acting as release planes. The red patches identify the joint Set 3, acting as a basal plane. Set 4, in green, is rarely mapped because of its geometrical orientation (near horizontal). Scale bar is indicative.

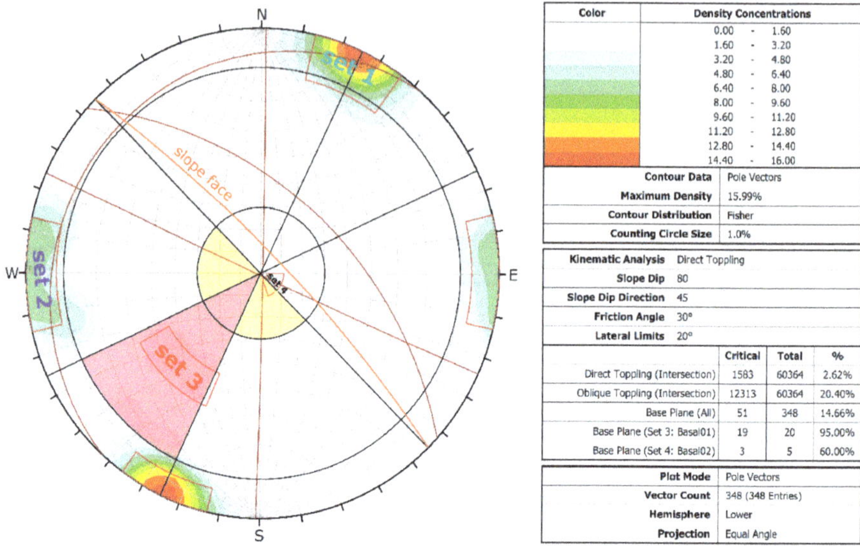

Color	Density Concentrations
	0.00 - 1.60
	1.60 - 3.20
	3.20 - 4.80
	4.80 - 6.40
	6.40 - 8.00
	8.00 - 9.60
	9.60 - 11.20
	11.20 - 12.80
	12.80 - 14.40
	14.40 - 16.00

Contour Data	Pole Vectors
Maximum Density	15.99%
Contour Distribution	Fisher
Counting Circle Size	1.0%

Kinematic Analysis	Direct Toppling
Slope Dip	80
Slope Dip Direction	45
Friction Angle	30°
Lateral Limits	20°

	Critical	Total	%
Direct Toppling (Intersection)	1583	60364	2.62%
Oblique Toppling (Intersection)	12313	60364	20.40%
Base Plane (All)	51	348	14.66%
Base Plane (Set 3: Basal01)	19	20	95.00%
Base Plane (Set 4: Basal02)	3	5	60.00%

Plot Mode	Pole Vectors
Vector Count	348 (348 Entries)
Hemisphere	Lower
Projection	Equal Angle

Figure 8. DIPS Stereoplot representation of the discontinuities extracted with SplitFX. The area shaded in red and yellow represents the area of instability highlighted by the kinematic analysis for direct and oblique toppling. The colour code associated with joint sets reflects the mapping performed on SplitFX in Figure 7.

Table 1. Summary of the geometrical orientation of the main joint sets identified in SplitFX.

Joint Set	Mean Dip (°)	Stdv (°)	Mean Dip Direction (°)	Stdv (°)
1	86.3	3.1	205.7	5.1
2	85.5	2.7	89.7	9.7
3	53.6	2.6	41.0	12.8
4	9.3	1.9	313.5	19.1

It can be seen from the kinematic analysis that the discontinuity mapping has highlighted the potential for both direct and oblique toppling from the slope, assuming a friction angle of 30°. In addition, the point cloud was used to establish typical block dimensions that could be formed by the respective discontinuity sets and be released in the event of rockfall. This was achieved by using the TLS point cloud and taking measurements perpendicular to the set orientation to obtain a true spacing and an estimate of the persistence [34]. The typical block size distribution comprises blocks ranging from 10 cm to 2 m in width. The rockfall deposit is scattered across an area of circa 430 m^2, and a visual comparison, aided by the measurement of blocks within the point cloud data, gives an estimated value for the total mobilised rock mass to be approximately 250 m^3.

In Figure 8 the result of the kinematic analysis for direct and oblique toppling are provided. The analysis is computed on 348 digitally mapped planar rock facets, assuming a general slope dip of 80°, the slope dip direction of N 45°, a standard friction angle of 30°, and lateral limits of 20°. The DIPS analysis shows how 20.24% of the discontinuities intersections potentially leading to oblique toppling fall into the instable area of the plot (shaded in red), justifying the assumed style of deformation for the case study.

3.2. 2D Rockfall Trajectory Analysis (Rocfall 2D-LM)

In this section the outcome of the 2D Lumped Mass rockfall trajectory analysis is presented. The input COR parameters used in the simulation are shown in Table 2. The COR were selected based on the literature available and the database values suggested by the software developer (literature value ± 3 stdv). Release points (or linear seeders) for the rockfall events were positioned at various locations at the top of the slope and along the sub-vertical rock wall. These source locations were validated through direct observation of relative fresh rockfall scars in the field and with the aid of high-resolution optical images (single close range frames from the drone survey). The overall number of rocks released from the seeders for each simulation run (5, 2, 1, 0.5 and 0.3 m resolution) was 10,000. Using the spacing and rock size distribution analysis performed previously, three rock classes, summarised below in Table 3, were selected for simulating different scenarios. The initial conditions of the blocks, in terms of initial horizontal and vertical velocity, were kept as default, i.e., zero angular and linear initial velocity.

Table 2. Summary of Coefficients of Restitutions (COR) used for the 2D-LM numerical modelling.

2D-LM Terrain Type	CoRN	CoRT	Friction Angle (°)
Granite/Rock face	Mean: 0.45 Std Dev: 0.04 Rel. Min: 0.12 Rel. Max: 0.12	Mean: 0.80 Std Dev: 0.04 Rel. Min: 0.12 Rel. Max: 0.12	Mean: 30 Std Dev: 0 Rel. Min: 0 Rel. Max: 0
Scree slope	Mean: 0.35 Std Dev: 0.04 Rel. Min: 0.12 Rel. Max: 0.12	Mean: 0.70 Std Dev: 0.03 Rel. Min: 0.9 Rel. Max: 0.9	Mean: 30 Std Dev: 0 Rel. Min: 0 Rel. Max: 0
Rock trap	Mean: 0.25 Std Dev: 0.04 Rel. Min: 0.12 Rel. Max: 0.12	Mean: 0.60 Std Dev: 0.04 Rel. Min: 0.12 Rel. Max: 0.12	Mean: 30 Std Dev: 0 Rel. Min: 0 Rel. Max: 0

Table 3. Summary of the rock classes defined for the 2D-LM numerical modelling.

2D-LM Rock Block Classes	Mass (kg)	Density (kg/m^3)
Small	Mean: 300 Std Dev: 25 Rel. Min: 75 Rel. Max: 75	Mean: 2650 Std Dev: 10 Rel. Min: 30 Rel. Max: 30
Medium	Mean: 1500 Std Dev: 50 Rel. Min: 150 Rel. Max: 150	Mean: 2650 Std Dev: 10 Rel. Min: 30 Rel. Max: 30
Large	Mean: 0.25 Std Dev: 0.04 Rel. Min: 0.12 Rel. Max: 0.12	Mean: 0.60 Std Dev: 0.04 Rel. Min: 0.12 Rel. Max: 0.12

The results of the 2D-LM rockfall analysis are shown in Figure 9. At the scale of this study, a relationship is observed between the geometrical resolution of the DEM and the distribution of rock paths end locations. The behaviour of the simulation, in terms of distribution of rock path end locations is influenced by the DEM resolution. In Figure 9c–e, showing 1, 2 and 5 m slope resolution, respectively, the results highlight uniform distributions, not consistent with the rock debris that can be observed in aerial pictures. As the DEM geometrical resolution increases up to 0.5 and 0.3 m (Figure 9a,b), the end locations distribution becomes more widespread along the slope and representative of the landslide body, as can be observed in Figure 1b. The visual comparison indicates how the majority of the smaller rock fragments are resting in the scree/transition zone mid-slope, and just the larger blocks reach the ditch at the base of the slope.

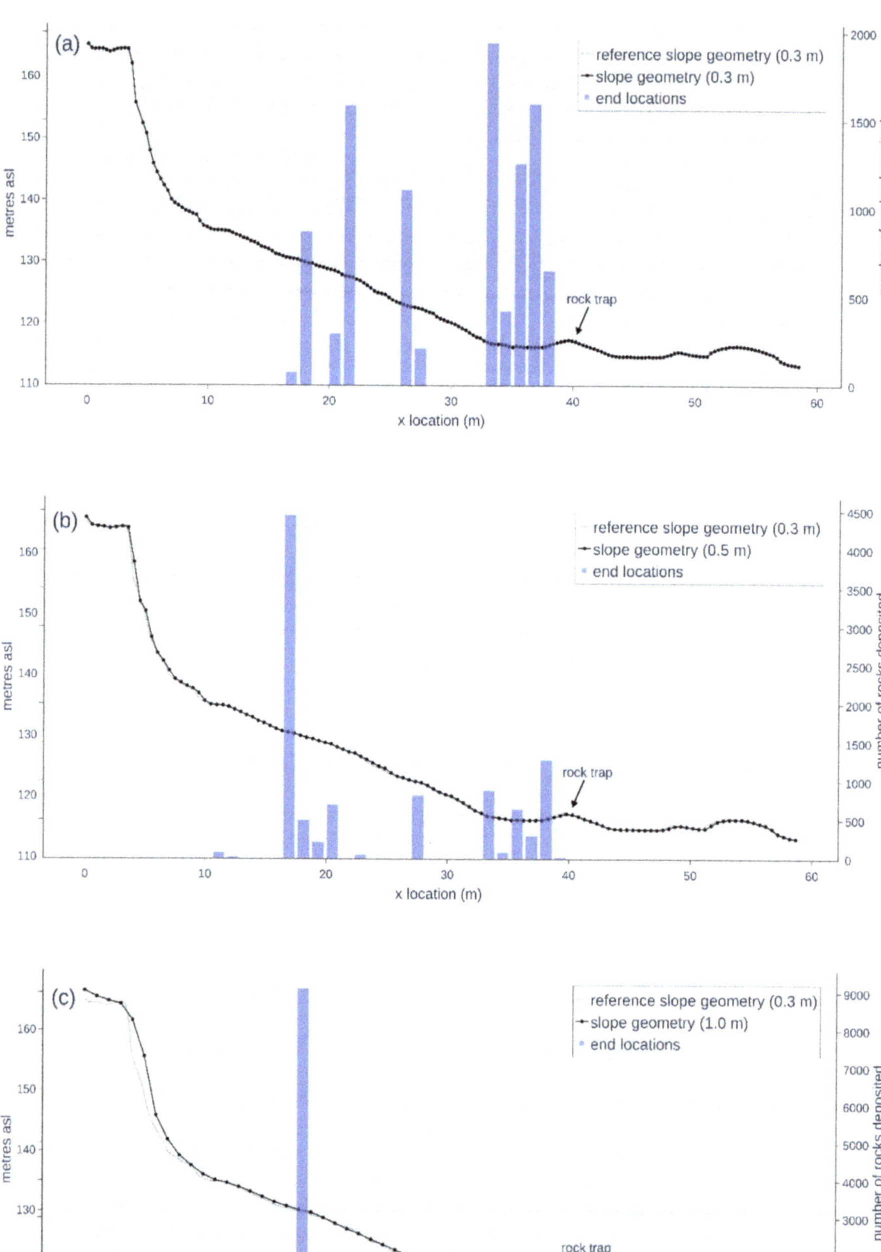

Figure 9. *Cont.*

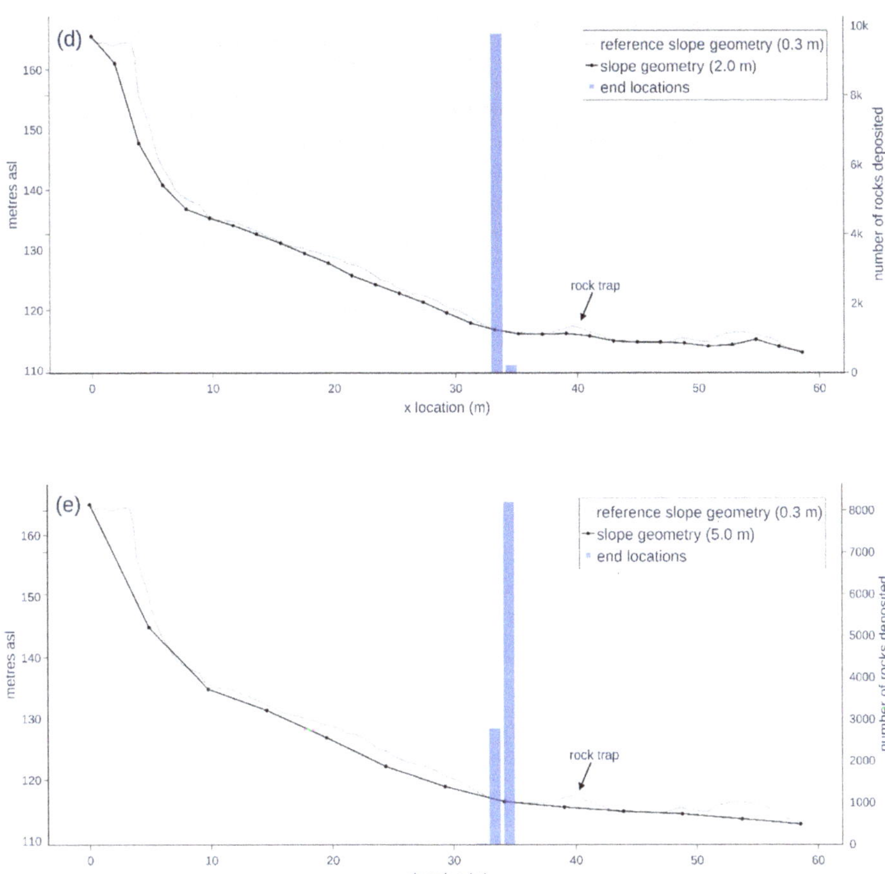

Figure 9. Results of the Rocfall 2D-LM. The plots show the distribution of end locations along the cross-section. The geometrical resolution adopted for each simulation run is as follows: (**a**) 0.3 m, (**b**) 0.5 m, (**c**) 1 m, (**d**) 2 m, (**e**) 5 m.

3.3. 3D Rockfall Trajectory Analysis (Rockyfor3D 3D-RB)

Rockyfor3D was used in order to assess the impact of a rockfall in three dimensions, rather than consider rockfall on a discrete 2D cross section of the slope. 3D analysis would also provide an insight into lateral dispersion of the rockfall deposit when propagating down the slope. The purpose of such simulation was to provide a spatial map of the distribution of end points of rock blocks trajectories, for a specific hazard scenario. The different scenarios were hypothesised based on the understanding of the rock mass conditions, obtained through the geo-mechanical analysis, as well as from screening historical aerial pictures and a geotechnical hazard assessment performed in the field with the help on the mining operation's personnel. The results are presented in the form of two GIS layers (i.e., reach probability and number of blocks deposited), that once combined, offer a statistically robust way to map the rockfall hazard to be used for risk assessment purposes. The validation phase was achieved through mapping end locations from previous events and comparing this with the results obtained from the simulation. This allows the user to calibrate the model against known end locations and their spatial distribution. The locations of release points were selected by applying an algorithm described in ARPA (2008) [45], which is based on the slope value (expressed in degrees) of each of the DEM's pixels. The algorithm sets a lower threshold depending on the DEM geometrical resolution;

every pixel in the DEM having a value exceeding that threshold is identified as a candidate release cell. After identification of the potential source locations, all the candidate pixels positions were screened and comparing pixel positions to aerial pictures. Where vegetation cover was present, the potential for release points was discounted. However, where the pixels coincided with exposed bare rock, they were included in the final selection as source areas or release points. As a result of running the ARPA's algorithm, 48 release points/seeders/pixels were extracted and used as the initial position for rock blocks in the 1 m resolution DEM (17 in the 2 m DEM and six in the 5 m DEM). For cell sizes smaller than 1 m the software reaches its computational limits and cannot compute any trajectory, hence those DEM having a resolution below 1 m were discarded. During each simulation run, selected as a combination of the DEM resolution (5, 2 and 1 m) and rock block classes (small, medium and large), a total of 10,000 block trajectories were simulated. The rock blocks volume, size and shape were set by extracting geometrical characteristics from the TLS/SfM point cloud (Figures 3 and 4). The initial velocity of falling blocks was simulated using a vertical freefall of 4 m (as can be observed in the annotated point cloud image in Figure 3, where the average vertical drop, from the height of the rockfall scars to the bottom of the vertical rock face is about 4–6 m). The COR values, summarised in Table 4, were attached to the raster layers by mapping areas of similar geotechnical properties within GIS. The maximum obstacle height (MOH), expressed as rg70%, rg20% and rg10% was representative of the obstacle height at the slope surface. This represents a statistical distribution of potential obstacles classes, whose values were determined by visual inspection of the slope.

Table 4. Summary of input parameters used for the three-dimensional rigid-body impact model (3D-RB) numerical modelling.

3D-RB Terrain Type	Rockyfor3D Soil Type	Mean CoRN	CoRN Value Range	Rg70 (m)	Rg20 (m)	Rg10 (m)
Vegetated slope	1—Fine soil material (depth > ~100 cm)	0.23	0.21–0.25	0.3	0.5	0.9
Scree slope	4—Talus slope (Ø > ~10 cm), or compact soil with large rock fragments	0.38	0.34–0.42	0.25	0.5	0.9
Rock trap	1—Fine soil material (depth > ~100 cm)	0.23	0.21–0.25	0.01	0.05	0.15
Haul road	5—Bedrock with thin weathered material or soil cover	0.43	0.39–0.47	0	0	0.1

The CORT is derived from this map through an implicit calculation of the software based on the statistical distribution of MOH values [44]. Figures 10–12 show the results of the simulations undertaken. These show the probability of a block to be arrested in a given cell of the DEM (reach probability) and the total number of blocks end location per pixel (number of blocks deposited). The simulations were run on DEM with different resolutions, namely 5 m (Figure 10), 2 m (Figure 11), and 1 m (Figure 12) and using the three rock classes, small, medium and large (Table 5), to observe the effect of the geometrical resolution of the DEM on modelled results, and the ability to capture fine scale irregularities within the topography.

Table 5. Summary of the rock classes defined for the 3D-RB numerical modelling.

3D-RB Rock Classes	Volume (m^3)	Block Shape (m)	Density (kg/m^3)	Mass (kg)
Small	0.125	Cubic 0.50 × 0.50 × 0.50	2650	331
Medium	0.576	Cubic 0.80 × 0.80 × 0.90	2650	1526
Medium	0.576	Rectangular 0.40 × 0.80 × 1.80	2650	1526
Large	1.000	Cubic 1.00 × 1.00 × 1.00	2650	2650

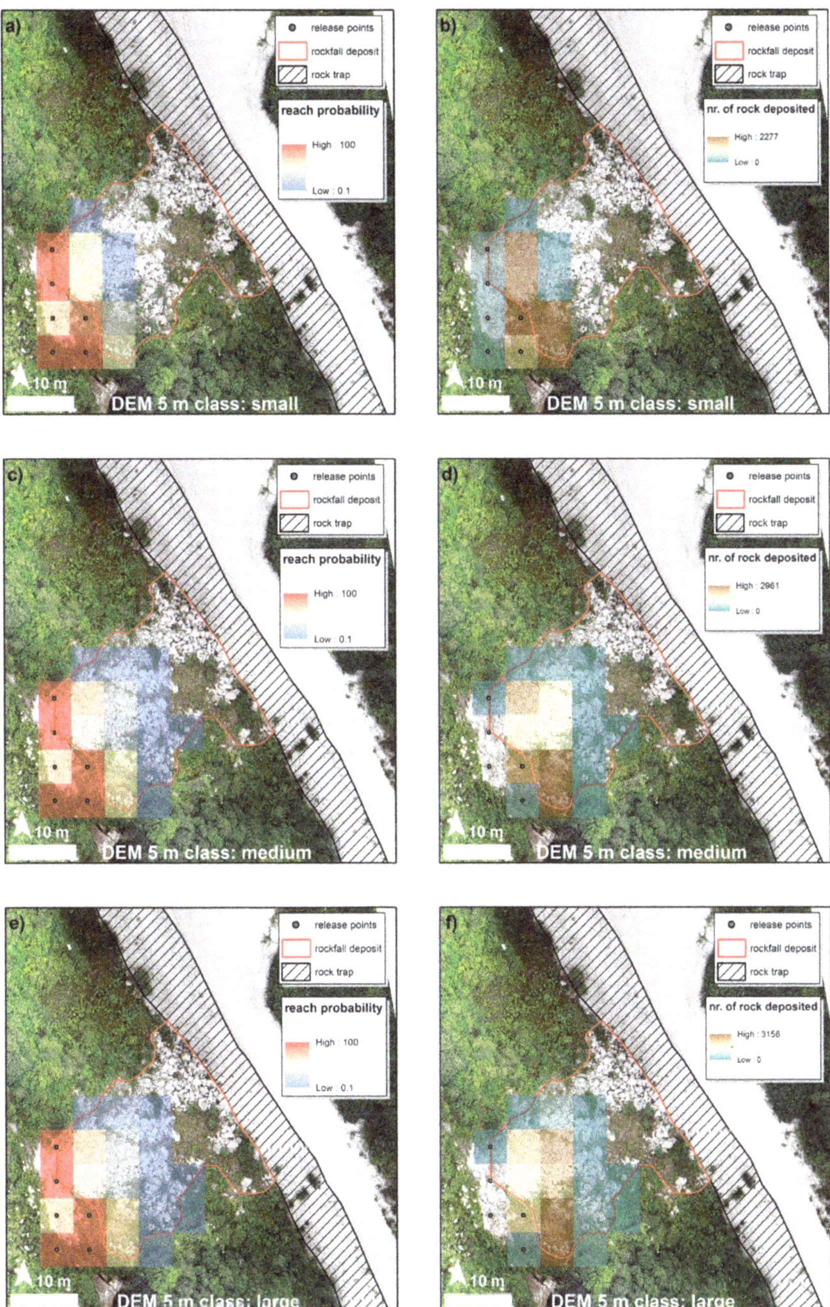

Figure 10. Rockyfor3D results computed on the 5 m resolution DEM. Left hand images (**a**,**c**,**e**) show the reach probability layers, on the right (**b**,**d**,**f**) the number of rocks deposited. The first row shows (**a**,**b**) results obtained with the rock class 'SMALL', second row (**c**,**d**) with 'MEDIUM' and the third row (**e**,**f**) with 'LARGE'. Rock block classes' properties are summarised in Table 5.

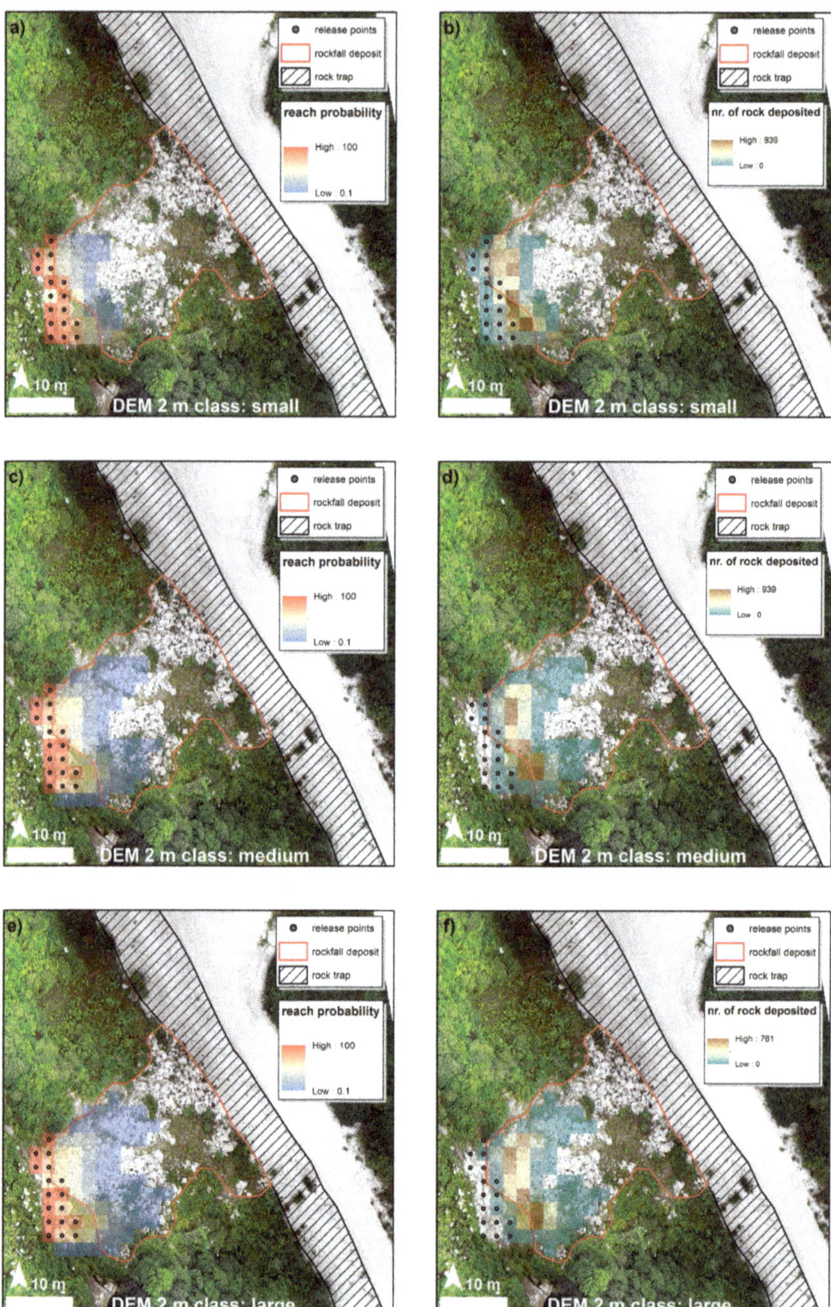

Figure 11. Rockyfor3D results computed on the 2 m resolution DEM. Left hand images (**a,c,e**) show the reach probability layers, on the right (**b,d,f**) the number of rocks deposited. The first row shows (**a,b**) results obtained with the rock class 'SMALL', second row (**c,d**) with 'MEDIUM' and the third row (**e,f**) with 'LARGE'. Rock block classes' properties are summarised in Table 5.

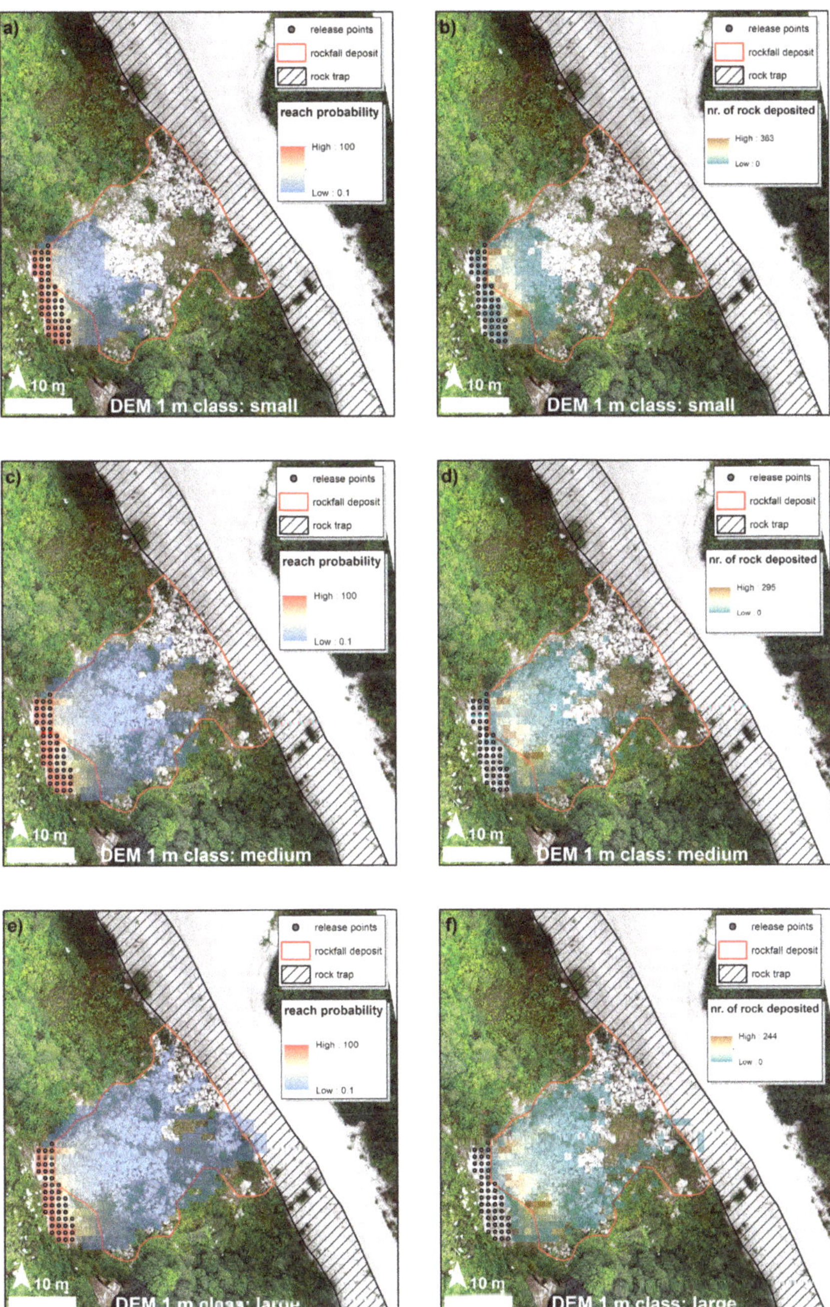

Figure 12. Rockyfor3D results computed on the 1 m resolution DEM. Left hand images (**a**,**c**,**e**) show the reach probability layers, on the right (**b**,**d**,**f**) the number of rocks deposited. The first row shows (**a**,**b**) results obtained with the rock class 'SMALL', second row (**c**,**d**) with 'MEDIUM' and the third row (**e**,**f**) with 'LARGE'. Rock block classes' properties are summarised in Table 5.

As part of the 3D modelling investigation, another variable was introduced: the rock block shape. Figure 13 shows the difference in terms of lateral spread and runout distance of blocks having the same volume and mass but a different shape, cubic in panel (a) and rectangular in panel (b). The increased reach of asymmetrical elongated blocks emerges for every topography resolution adopted.

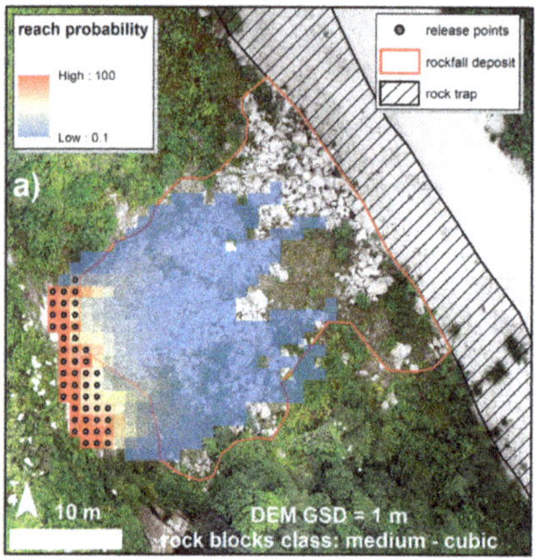

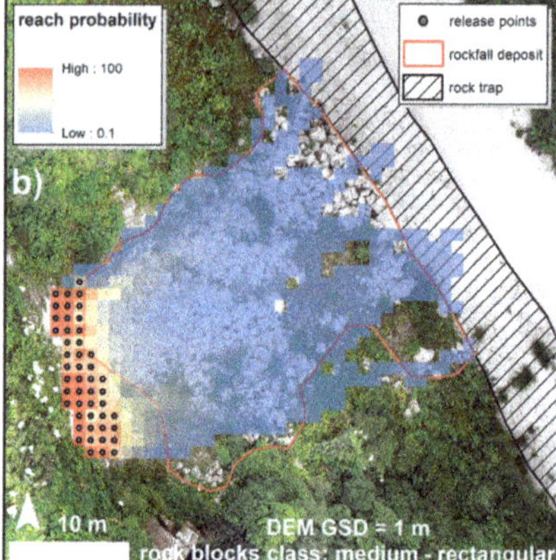

Figure 13. Comparison of reach probability layers showing the effect of block shape; panel (**a**) shows the runout of equidimensional (cubic) blocks, while panel (**b**) shows the runout of elongated (rectangular) blocks. Rock block classes' properties are summarised in Table 5.

4. Discussion

The results obtained from the rockfall trajectory analysis have provided insights into the behaviour of the rockfall event(s), while exploring the effectiveness of remotely sensed data (ALS, TLS and SfM derived point clouds) as basis for creating DEM for numerical modelling. The reach probability maps and the distribution of end points obtained with both the 2D-LM and 3D-RB approaches showed how there is a positive correlation between the resolution of the DEM and the simulated trajectories. Higher resolution DEM are capable of capturing small scale irregularities, resulting in an increased variability of the end locations. DEM up to 1 m GSD were obtained from ALS surveys, while very high resolution DEM (GSD ≤ 1 m) were obtained from either TLS or images acquired with a UAV. The moderate resolution DEM, when implemented in the simulations, gave rise to slower and less energetic trajectories. This analysis would suggest the controlling influence of slope resolution geometry on modelled rockfall end point location.

It is known that large roll out distances are possible when a falling rock's translational momentum is changed into rotational momentum by impacting the slope, and that launch features may change a rock's vertical drop to horizontal displacement [4,46]. This analysis confirms that back analysis of rockfall events provide an opportunity to investigate the impact of the geometrical parameters influencing the roll out distance and the distribution of rock paths end locations. The 3D analyses show an acceptable visual correlation between the geometry of the landslide body and the reach probability maps obtained, both in terms of the spread and runout, as shown in Figure 12c–f. The 3D simulation was able to effectively capture the observed typical rockfall trajectory and lateral extent of the resultant rockfall debris.

The 3D modelling has highlighted the impact of the Maximum Obstacle Height (MOH) on results. It is therefore important to undertake sensitivity analyses on this key input parameter during back analyses and establish statistically robust distributions in 3D-RB rockfall simulation using RF3D. The study has highlighted that, where possible, it is important to include field mapping to provide rigorous validation of data. Results from modelling undertaken on this case study demonstrate the dilemma in rockfall simulation. Unrealistic results are obtained where the surface topography included in the model is too coarse; unrealistic behaviour of modelled rockfall trajectories can also arise when inappropriate COR are used as inputs. From the analyses undertaken there is an inability of the three-dimensional model to correctly simulate observed rockfall trajectories for high resolution DEM which results from poor fine scale mapping, used to associate input parameters to the digital topography, and results in poor zonation of end locations. Dorren (2004) [47] suggests the use of DEM ranging from 25 to 2 m, for regional scale and slope scale simulations. This case study has highlighted considerable variance in rockfall trajectory when increasing the DEM resolution, up until reaching the computational limits of the software and CPU. It is important therefore to calibrate rock block classes, COR, and MOH classes and include a topography of a specific resolution that enables a robust representation of specific scenario being modelled. The resolution achieved with both TLS and SfM is considerably greater than the one selected to generate DEM at the grid sizes used in this study, so it appears that the computational power/hardware requirements/algorithms remain the main obstacle to the use of sub-metric DEM. It is important to acknowledge the impact of DEM on modelled results and the need for guidelines to target optimal resolutions when generating DEM to be used for rockfall trajectory analyses.

As part of the modelling undertaken, another key aspect was the definition of the rock-trap geometry used within the models. The barrier in place to restrict the horizontal travel distance of rock blocks is a critical part of the rock-trap system. However, given its geometry (i.e., an embankment, usually made of sand or crushed rocks), the inner side of the embankment is occluded from a position external to the rock-trap itself, such as the haul road. This condition is usually overcome by capturing the scene from a mobile platform, such as a UAV. The ALS derived model proved to be useful when used in numerical modelling, provided they did not include any region of occlusion. The steepness of the rock wall represents an unfavourable condition that can be addressed by ALS with due precautions (i.e., a careful flight plan design, so to avoid occlusions while capturing the scene), but this is not often feasible, as for this study the ALS dataset was obtained without this specific need in mind, resulting in some data useless for the purpose of numerical modelling. TLS derived models were also ineffective as they were unable to image the inner zone of the rock-trap. From Figure 14 it appears that, regardless of the RS technique used, the 1 m resolution DEM represents the maximum threshold necessary for obtaining an optimal description of the rock-trap geometry.

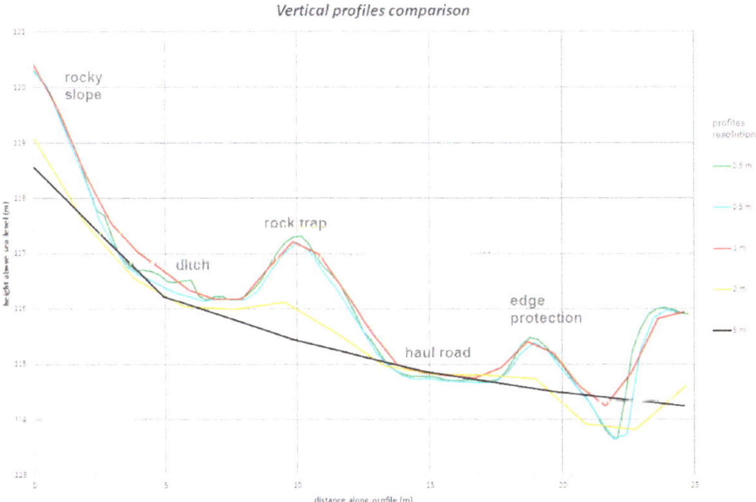

Figure 14. Comparison of vertical profiles, extracted from the DEM at different resolutions. The vertical axis is exaggerated by a factor of two.

From the analyses undertaken, the shape of blocks modelled also has a significant influence on the rockfall simulation results. In the 3D-RB approach, the medium rectangular class appears to reach further distances compared to equidimensional (cubic) blocks having same mass and volume, going against the general understanding that larger mass and inertia will lead to a longer runout. Although it is not clear what mechanism adopted in the models could lead to such an outcome, the main hypothesis is that elongated blocks tend to roll perpendicular to their major axis, hence gaining high angular momentum compared to the equidimensional blocks. This increased angular velocity is directly linked to a greater horizontal travel distance for rock fragments.

The results show the importance of undertaking both two-dimensional and three-dimensional modelling for the case study, but emphasis is given for the need of both calibration and validation of results to ensure confidence for future use in hazard and risk evaluation.

5. Conclusions

The case study highlights how 3D photogrammetric remotely sensed data can be effectively used to inform the back analysis of a rockfall event at slope scale. The topographic reconstruction of the 3D scene was obtained by using different remote mapping techniques which included aerial and terrestrial laser scanning, and image acquisition by UAV. The resultant high resolution point clouds were then analysed by extracting geo-mechanical features to characterise the rockmass. This included definition of the orientation and spatial distribution of discontinuities within the rock slope. The discontinuity data was subsequently used to perform kinematic admissibility analysis to highlight the potential for both direct and oblique toppling from the columnar jointing. The point cloud data was also used to establish in-situ block size and rockfall block size distribution from the rockfall debris for input data and validation respectively.

Analysis of a series of aerial images was used to determine evolution phases of the rockfall and establish the size and spatial distribution of rock blocks resting on the slope (end point locations). Geo-mechanical and geotechnical features were then translated into modelling parameters, to allow a probabilistic, process-based rockfall trajectory analysis to be performed using both two- and three-dimensional approaches.

The results of the three-dimensional modelling show how the modelling can realistically capture rockfall trajectories in terms of spatial distribution and runout pathways. The models were able to demonstrate the effectiveness of the existing rock-trap. However, the results show the importance and need for calibration of input parameters, as modelled results are clearly influenced by the resolution of the surface topography used within the models. Validation of models through comparison with end point locations is therefore essential for confidence in future use of such models for hazard and risk assessment. The results of the analysis would suggest that guidelines are necessary when using remote mapping data for generation of surface topography for rockfall trajectory analysis, as the spatial resolution of the surface topography has a critical influence on the modelled behaviour. Guidelines are therefore needed to establish suitable DEM resolutions for generation of surface topographies that enable realistic rockfall simulation.

The ability to assess the rockfall behaviour in both two and three dimensions greatly improves the understanding of hazards posed to the mining operation. The methodology proposed within this study can provide the basis for calibration of rockfall input parameters relevant to the case study site and therefore provide the framework for future rockfall hazard assessment and evaluation. This will then provide the basis for risk assessment and design of suitable protection measures.

Author Contributions: Investigation, Carlo Robiati, Matt Eyre, Mirko Francioni, John Coggan and Adam Venn; Formal Analysis, Carlo Robiati, Matt Eyre and John Coggan; Data Curation, Carlo Robiati, Matt Eyre and John Coggan; Methodology and Validation, Carlo Robiati, Matt Eyre, Mirko Francioni and John Coggan; Supervision, Mat Eyre and John Coggan; Writing-original draft, Carlo Robiati; Writing-review and editing, Carlo Robiati, Matt Eyre, Claudio Vanneschi, Mirko Francioni and John Coggan.

Funding: This research received no external funding.

Conflicts of Interest: The authors declare no conflict of interest.

References

1. Riquelme, A.; Cano, M.; Tomás, R.; Abellán, A. Identification of Rock Slope Discontinuity Sets from Laser Scanner and Photogrammetric Point Clouds: A Comparative Analysis. *Procedia Eng.* **2017**, *191*, 838–845. [CrossRef]
2. *Health and Safety at Quarries. Quarries Regulations 1999. Approved Code of Practice and Guidance*; L118 (Second edition); Health and Safety Executive: London, UK, 2013.
3. Hutchinson, J.N. General Report: Morphological and geotechnical parameters of landslides in relation to geology and hydrogeology. In Proceedings of the Fifth International Symposium on Landslides, Lausanne, Switzerland, 10–15 July 1988; Bonnard, C., Ed.; Balkema: Rotterdam, The Netherlands, 1988; Volume 1, pp. 3–35.
4. Ritchie, A.M. *Evaluation of Rockfall and Its Control*; Washington State Highway Commission, Committee on Landslide Investigations: Olympia, WA, USA, 1961.
5. Varnes, D.J. Slope Movements. Types and Processes. TRB Special Report 176. In *Landslides: Analysis and Control*; Transportation Research Board: Washington, DC, USA, 1978; pp. 11–33.
6. Cruden, D.M.; Varnes, D.J. *Landslide Types and Processes*; Special Report; Transportation Research Board, U.S. National Academy of Sciences: Washington, DC, USA, 1996; Volume 247, pp. 36–75.
7. McCauley, M.L.; Works, C.B.W.; Naramore, S.A. *Rockfall Mitigation*; California Department of Transportation Report: Sacramento, CA, USA, 1985.
8. Wyllie, D.C.; Mah, C.W. *Rock Slope Engineering: Civil and Mining*, 4th ed.; Routledge: London, UK, 2004; p. 456.
9. Chau, K.T.; Wong, R.H.C.; Lee, C.F. Rockfall Problems in Hong Kong and Some New Experimental Results for Coefficients of Restitution. *Int. J. Rock Mech. Min. Sci.* **1998**, *35*, 662–663. [CrossRef]
10. Chau, K.T.; Wong, R.H.C.; Wu, J.J. Coefficient of restitution and rotational motions of rockfall impacts. *Int. J. Rock Mech. Min. Sci.* **2002**, *39*, 69–77. [CrossRef]
11. Crosta, G.B.; Agliardi, F. A methodology for physically based rockfall hazard assessment. *Nat. Hazards Earth Syst. Sci.* **2003**, *3*, 407–422. [CrossRef]
12. Dorren, L.K.A. A review of rockfall mechanics and modelling approaches. *Prog. Phys. Geogr.* **2003**, *27*, 69–87. [CrossRef]
13. Zhou, Y. Application of rockfall risk assessment techniques in two aggregate quarries. In *Harmonising Rock Engineering and the Environment*; Zhou, Y. CRC Press: Boca Raton, FL, USA, 2011; pp. 1861–1864, ISBN 978-0-415-80444-8.
14. Corominas, J.; Matas, G.; Ruiz-Carulla, R. Quantitative analysis of risk from fragmental rockfalls. *Landslides* **2019**, *16*, 5–21. [CrossRef]
15. Jaboyedoff, M.; Dudt, J.P.; Labiouse, V. An attempt to refine rockfall hazard zoning based on the kinetic energy, frequency and fragmentation degree. *Nat. Hazards Earth Syst. Sci.* **2005**, *5*, 621–632. [CrossRef]
16. Salvini, R.; Francioni, M.; Riccucci, S.; Bonciani, F.; Callegari, I. Photogrammetry and laser scanning for analysing slope stability and rock fall runout along the Domodossola-Iselle railway, the Italian Alps. *Geomorphology* **2013**, *185*, 110–122. [CrossRef]
17. Abellán, A.; Vilaplana, J.M.; Martínez, J. Application of a long-range Terrestrial Laser Scanner to a detailed rockfall study at Vall de Núria (Eastern Pyrenees, Spain). *Eng. Geol.* **2006**, *88*, 136–148. [CrossRef]
18. Francioni, M.; Stead, D.; Clague, J.J.; Westin, A. Identification and analysis of large paleo-landslides at Mount Burnaby, British Columbia. *Environ. Eng. Geosci.* **2018**, *24*, 221–235. [CrossRef]
19. Francioni, M.; Coggan, J.; Eyre, M.; Stead, D. A combined field/remote sensing approach for characterizing landslide risk in coastal areas. *Int. J. Appl. Earth Obs. Geoinf.* **2018**, *67*, 79–95. [CrossRef]
20. Jaboyedoff, M.; Oppikofer, T.; Abellán, A.; Derron, M.-H.; Loye, A.; Metzger, R.; Pedrazzini, A. Use of LIDAR in landslide investigations: A review. *Nat. Hazards* **2012**, *61*, 5–28. [CrossRef]
21. Lato, M.J.; Vöge, M. Automated mapping of rock discontinuities in 3D lidar and photogrammetry models. *Int. J. Rock Mech. Min. Sci.* **2012**, *54*, 150–158. [CrossRef]

22. Fukuzono, T. A new method for predicting the failure time of a slope. Proceedings of 4th International Conference and Field Workshop on Landslides, Tokyo, Japan, 23–31 August 1985; pp. 145–150.

23. Hutchinson, D.J.; Lato, M.; Gauthier, D.; Kromer, R.; Ondercin, M.; van Veen, M.; Harrap, R. Applications of remote sensing techniques to managing rock slope instability risk. In Proceedings of the GeoQuebec 2015, Quebec City, QC, Canada, 20–23 September 2015; p. 11.

24. Martino, S.; Mazzanti, P. Integrating geomechanical surveys and remote sensing for sea cliff slope stability analysis: The Mt. Pucci case study (Italy). *Nat. Hazards Earth Syst. Sci.* **2014**, *14*, 831–848. [CrossRef]

25. Salvini, R.; Francioni, M.; Riccucci, S.; Fantozzi, P.L.; Bonciani, F.; Mancini, S. Stability analysis of "Grotta delle Felci" Cliff (Capri Island, Italy): Structural, engineering–geological, photogrammetric surveys and laser scanning. *Bull. Eng. Geol. Environ.* **2011**, *70*, 549–557. [CrossRef]

26. Tonini, M.; Abellan, A. Rockfall detection from terrestrial LiDAR point clouds: A clustering approach using R. *J. Spat. Inf. Sci.* **2014**, *8*, 95–110. [CrossRef]

27. Carrivick, J.L.; Smith, M.V.; Quincey, D.J. *Structure from Motion in the Geosciences*; Wiley Blackwell: New York, NY, USA, 2016; p. 210.

28. Francioni, M.; Salvini, R.; Stead, D.; Coggan, J. Improvements in the integration of remote sensing and rock slope modelling. *Nat. Hazards* **2018**, *90*, 975–1004. [CrossRef]

29. Giordan, D.; Hayakawa, Y.; Nex, F.; Remondino, F.; Tarolli, P. Review article: The use of remotely piloted aircraft systems (RPASs) for natural hazards monitoring and management. *Nat. Hazards Earth Syst. Sci.* **2018**, *18*, 1079–1096. [CrossRef]

30. Salvini, R.; Mastrorocco, G.; Esposito, G.; Di Bartolo, S.; Coggan, J.; Vanneschi, C. Use of a remotely piloted aircraft system for hazard assessment in a rocky mining area (Lucca, Italy). *Nat. Hazards Earth Syst. Sci.* **2018**, *18*, 287–302. [CrossRef]

31. Ellis, R.J.; Scott, P.W. Evaluation of hyperspectral remote sensing as a means of environmental monitoring in the St. Austell China clay (kaolin) region, Cornwall, UK. *Remote Sens. Environ.* **2004**, *93*, 118–130. [CrossRef]

32. Hill, P.I.; Howe, J.H. Primary lithological variation in the kaolinized St Austell Granite, Cornwall, England. *J. Geol. Soc.* **1996**, *153*, 827–838. [CrossRef]

33. Tannant, D. Review of Photogrammetry-Based Techniques for Characterization and Hazard Assessment of Rock Faces. *Int. J. Geohazards Environ.* **2015**, *1*, 76–87. [CrossRef]

34. Sturzenegger, M.; Stead, D. Quantifying discontinuity orientation and persistence on high mountain rock slopes and large landslides using terrestrial remote sensing techniques. *Nat. Hazards Earth Syst. Sci.* **2009**, *9*, 267–287. [CrossRef]

35. Markland, J.T. *A Useful Technique for Estimating the Stability of Rock Slopes when the Rigid Wedge Slide Type of Failure is Expected*; Rock mechanics research report; report no. 19; Interdepartmental Rock Mechanics Project, Imperial College of Science and Technology: London, UK, 1972.

36. Turner, A.K.; Duffy, J.D. Modelling and prediction of Rockfalls. In *Rockfall: Characterization and Control*; TRB: Washington, DC, USA, 2012.

37. Li, L.; Lan, H. Probabilistic modelling of rockfall trajectories: A review. *Bull. Eng. Geol. Environ.* **2015**, *74*, 1163–1176. [CrossRef]

38. Asteriou, P.; Saroglou, H.; Tsiambaos, G. Geotechnical and kinematic parameters affecting the coefficients of restitution for rock fall analysis. *Int. J. Rock Mech. Min. Sci.* **2012**, *54*, 103–113. [CrossRef]

39. Bar, N.; Nicoll, S.; Pothitos, F. Rock fall trajectory field testing, model simulations and considerations for steep slope design in hard rock. In Proceedings of the First Asia Pacific Slope Stability in Mining Conference, Brisbane City, Australia, 6–8 September 2016; p. 10.

40. Azzoni, A.; Rossi, P.P.; Drigo, E.; Giani, G.P.; Zaninetti, A. In situ observation of rockfall analysis parameters. In Proceedings of the VI International Symposium on Landslides, Christchurch, South Island, New Zealand, 10–14 February 1992.

41. Bourrier, F.; Berger, F.; Tardif, P.; Dorren, L.; Hungr, O. Rockfall rebound: Comparison of detailed field experiments and alternative modelling approaches. *Earth Surf. Process. Landf.* **2012**, *37*, 656–665. [CrossRef]

42. Spadari, M.; Giacomini, A.; Buzzi, O.; Fityus, S.; Giani, G.P. In situ rockfall testing in New South Wales, Australia. *Int. J. Rock Mech. Min. Sci.* **2012**, *49*, 84–93. [CrossRef]

43. Stevens, W.D. ROCFALL: A Tool for Probabilistic Analysis, Design of Remedial Measures and Prediction of Rockfalls. Master's Thesis, University of Toronto, Toronto, ON, Canada, 1988.

44. Dorren, L.K.A. Rockyfor3D v5.2 revealed. In *Transparent Description of the Complete 3D Rockfall Model*; ecorisQ paper; ecorisQ: Geneva, Switzerland, 2016; p. 32.
45. Arpa Piemonte—Centro Regionale per le Ricerche Territoriali e Geologiche. *ARPA Progetto n. 165 PROVIALP, Protezione Della Viabilità Alpine—Relazione Finale*; Arpa Piemonte: Milan, Italy, 2008.
46. Pierson, L.A.; Gullixson, C.F.; Chassie, R.G. *Rockfall Catchment Area Design Guide*; Final Report (Metric Edition); Oregon Department of Transportation and Federal Highway Administration: Washington, DC, USA, 2001.
47. Dorren, L.K.A.; Maier, B.; Putters, U.S.; Seijmonsbergen, A.C. Combining field and modelling techniques to assess rockfall dynamics on a protection forest hillslope in the European Alps. *Geomorphology* **2004**, *57*, 151–167. [CrossRef]

 International Journal of
Geo-Information

Article

SfM-MVS Photogrammetry for Rockfall Analysis and Hazard Assessment Along the Ancient Roman Via Flaminia Road at the Furlo Gorge (Italy)

Claudio Vanneschi [1,*], Marco Di Camillo [2], Eros Aiello [3], Filippo Bonciani [3] and Riccardo Salvini [4]

[1]   CGT Spinoff s.r.l., Via Vetri Vecchi 34, 52027 San Giovanni Valdarno (AR), Italy
[2]   Terna Rete Italia S.p.A., Viale Egidio Galbani 70, 00156 Roma, Italy
[3]   CGT Engineering S.r.l., Via Strasburgo 7, 52022 Cavriglia (AR), Italy
[4]   Department of Environment, Earth and Physical Sciences and Centre for GeoTechnologies CGT, University of Siena, Via Vetri Vecchi 34, 52027 San Giovanni Valdarno (AR), Italy
*   Correspondence: vanneschi@cgt-spinoff.it

Received: 3 July 2019; Accepted: 24 July 2019; Published: 25 July 2019

Abstract: Rockfall events represent significant hazards for areas characterized by high and steep slopes and therefore effective mitigation controls are essential to control their effect. There are a lot of examples all over the world of anthropic areas at risk because of their proximity to a rock slope. A rockfall runout analysis is a typical 3D problem, but for many years, because of the lack of specific software, powerful computers, and economic reasons, a 2D approach was normally adopted. However, in recent years the use of 3D software has become quite widespread and different runout working approaches have been developed. The contribution and potential use of photogrammetry in this context is undoubtedly great. This paper describes the application of a 3D hybrid working approach, which considers the integrated use of traditional geological methods, Terrestrial Laser Scanning, and drone based Digital Photogrammetry. Such approach was undertaken in order to perform the study of rockfall runout and geological hazard in a natural slope in Italy in correspondence of an archaeological area. Results show the rockfall hazard in the study area and highlights the importance of using photogrammetry for the correct and complete geometrical reconstruction of slope, joints, and block geometries, which is essential for the analysis and design of proper remediation measures.

Keywords: unmanned aerial vehicle; photogrammetry; rock slope stability; rockfall runout; geological hazard

---

1. Introduction

Rockfall is a common natural process acting on rock slopes. Such phenomenon is usually characterized by unpredictability and high velocities, representing a significant hazard especially near communication routes, villages, workplaces (for instance either open pit mines or dams), and other popular areas (such as archaeological, historical, or religious sites). A rockfall event can be very dangerous, causing casualties even when a relatively small mass is mobilized [1,2], and can be triggered by one or a combination of different factors, such as climate, seismic activity, and vegetation growth. These factors can cause the opening of joints, pore pressure increase, freeze-thaw processes, or chemical weathering of rock [3]. Once a rockfall has occurred, the trajectory and the energy of falling blocks are controlled by a complex interaction of multiple factors including the shape and size of the block, micro- and macro-geometry of the slope, and its surface properties (e.g., roughness, material properties, land cover). Hence, it is important to study the possible trajectory and energy of a rockfall since often the related hazards cannot be eliminated because the magnitude and frequency of rockfall events can vary both spatially and temporally [4].

Rockfall can be analyzed either through 2D [5–8] or more recently using 3D analyses [9–15].

For many years the analysis of a rockfall event was limited to bi-dimensional simulations due to the lack of detail in the Digital Terrain Model (DTM) and computing constraints (both software and hardware related). 2D rockfall simulations have been performed and considered statistically very significant; this was mainly due to the larger number of trajectories that could be elaborated and analyzed [16]. However, with new efficient software, capable of performing 3D runout analyses without considerable computing power, applications are being explored in the scientific community [14,17–19]. The results of rockfall analyses are represented, among the others, by runout distances, velocities, kinetic energies, probability, and bounce heights of rock blocks. In general, a limiting factor for both 2D and 3D approaches may be represented by the lack of high-resolution geometrical information of the slope surface and the falling block, especially in 3D analyses [20–22]. The difficulties in obtaining accurate geometrical information of the falling block and trajectories can result in a high level of uncertainty, which requires the need of probabilistic approaches [6,23]. In general, when dealing with uncertainty, statistics represent a way to provide useful answers for intervention planning, since it leads to a precautionary approach: this, for example, is typical of slope stability analysis, where sensitivity or probabilistic approaches are undertaken on strength and stress values in order to assess their influence on the Safety Factor (SF).

Traditional survey methods and topographic maps are typically used to model slopes. However, such techniques have important limitations in collecting the necessary spatial data required for proper runout simulations. The scale of the most used topographic maps (typically 1:1000, 1:2000, or smaller) and related accuracy (between 0.4 and 1.0 m or greater), together with the use of the nominal size of falling blocks, may lead to the inaccurate estimations of runout distances, velocities, kinetic energies, and bounce heights. To enable topographic representation and element sizing, Geomatic techniques can be adopted for detailed and accurate data acquisitions [24–28]. In fact, Geomatics is a discipline that deals, among others, with acquiring, modeling, interpreting, and processing scaled and geo-referenced information. Geomatics has benefited from the recent development of information technology and sensors, which improved the acquisition and processing of geometrical information and the monitoring of infrastructure and territory from geological risks. Geomatics includes terrestrial positioning techniques (historically included in the Topography) and spatial positioning techniques (e.g., Global Navigation Satellite System—GNSS), Digital Photogrammetry (DP), laser scanning techniques, remote sensing, Geographic Information Systems (GIS, including WebGIS and geoservices), and geostatistics.

Laser scanning is an optimized technique to acquire 3D topographic data [29–34]. Laser scanning can be utilized terrestrially (TLS) to record near vertical structures or Airborne (ALS) to map larger areas. Typically, TLS has a spatial resolution ranging from millimeters to centimeters, while ALS, acquire data at lower ground sampling distances, from centimeter to decimeter [35–37]. Even DP allows centimetric representations of the topography of an area [38–42]. In regard to this, the widespread use of Unmanned Aerial Vehicles (UAVs), together with Structure from Motion (SfM, [43]) and Multi-View Stereo (MVS, e.g., [44–46]) technologies has improved the flexibility of DP. In fact, DP can nowadays easily be used under different conditions: nadiral, oblique, and frontal images can be acquired with no particular effort. In this context the presence of vegetation is still a crucial factor and is more relevant in photogrammetry due to optical data collection that cannot penetrate vegetated areas as with full waveform laser scanning utilizing multi-echo returns.

With the advancement in remote sensing techniques, several researchers have successfully applied LS or DP in rockfall analyses [47–56]. When dealing with rockfall runout analyses, different kinds of approaches are generally adopted. As indicated by [57], the simplest methods refer to empirical rating systems that are based on the use of historical catalogues of past rockfall events [1,58]. Other methods take advantage from a deterministic approach, calculating the SF of potentially unstable blocks by means of limit-equilibrium methods. Alternatively, considering that in many situations the rockfall hazard varies both spatially and temporally, statistical based approaches are often preferred [11]. In this paper, a hybrid approach that combines both deterministic and statistical methods was adopted to perform 3D runout analysis of a natural rock slope in Italy (Figure 1). The site, where some rockfall events have already occurred in the past, was analyzed because of the risk posed by the presence of the local ancient Roman Via Flaminia road, close to an important archaeological site of Roman age, and the Church named "Santa Maria delle Grazie." A specific multi-source survey, using a Total Station (TS), GNSS, TLS, and UAV, was performed to characterize the steep rock slope (practically inaccessible without mountaineering-climbing equipment and relative capacity) and determine the location and geometry of potentially unstable blocks visually identified by photointerpretation. The collected data was then utilized to perform specific slope stability studies, 3D runout analyses, and to produce a new rockfall hazard map. Specifically, the aim of this study was to make available to professionals of Enel Green Power O&M Hydro Italy with hydro-civil function of the Furlo dam, a deterministic analysis containing rock block volumes, weights, stability SFs, and runout analyses to be used as quantitative data for protection measures planning.

Figure 1. Location of the study area (**a**): aerial orthophoto of "Canavaccio" Section (Nr. 280090 from the Marche Regional Technical Cartography at a scale of 1:10,000) used as background image; inset map shows the location of the study area in Italy; perspective view of the Furlo Gorge from an Unmanned Aerial Vehicle (UAV) (**b**): in the background the dam, in the foreground the Roman Via Flaminia road (to the left), and the Church of "Santa Maria delle Grazie" (in the center).

2. Study Area

The study area is in the Municipality of Fermignano, Province of Pesaro-Urbino. The Furlo Gorge has developed between the Pietralata Mount (889 m a.s.l.) and the Paganuccio Mount (976 m a.s.l.) due to the erosional power of the Candigliano River; during the years the river has reached a considerable depth in this area, which is currently no longer visible due to the presence of the dam, built in 1922, which reduced the impetuous river to a placid lake (Figure 1).

In Roman times, with the aim of connecting people in such an impervious area, the Emperor Vespasian dug a tunnel in the narrowest point of the gorge that was called "petra pertusa" or "forulum" (small hole), from which the name "Furlo" was generated. Next to the Furlo, there is an earlier Etruscan passageway, 8 m long, 3.30 m wide, and 4.45 m high, and a small Church, also called "della Botte," once inhabited by a hermit. The tunnel was completely carved into the rock by a chisel, in which graded cuts are still visible.

The slopes are made up of limestone of marine origin belonging to the Mesozoic Umbria-Marche Succession of the Italian Apennine [59]. Limestones mainly include the Calcare Massiccio Formation dated to Early Jurassic (i.e., Hettangian-Early Sinemurian stages). This formation is characterized by the absence of stratification with large banks typical of carbonate platform environment but presents numerous high-angle fault systems and discontinuities, both tectonic and arising from unloading [60–62]. The color of the Calcare Massiccio Formation is typically white or greyish, while its thickness reaches up to 300 m [59].

From a structural-geological perspective, the study area is located within a wide anticline of Apennine origin which has at the core the Calcare Massiccio Formation; the compressive stresses responsible of the Apennine chain formation acting since at late Paleogene period (i.e., late Oligocene epoch) generated a NW-SE oriented anticline with a sub-vertical axial plane, sub-horizontal axis, and an open cylindrical geometry [61].

The core shows a collapse structure identified by two direct faults NW-SE oriented that lower the central portion of the fold and determine the formation of an elongated graben along the direction of the axial plane. The fault displacements are of a few tens of meters. The anticline is an uprooted structure, translated to NE above a basal thrust which is temporally extended towards NE, from the bottom upwards, involving more recent stratigraphic intervals in the deformation [63]. The anticline shows a radius of curvature greater than 2 km and is characterized by a symmetry with orientation N 125–130 and a barely noticeable gradient towards NE.

In the Furlo Gorge collapses are extremely widespread with several sub-vertical fronts of various origins. Slope detachments occur very often due to physical-chemical weathering along stratification beds, networks of tectonic and/or newly formed joints produced by unloading processes [64].

Previous research on rockfalls at the Furlo Gorge report of past events that often affected the Via Flaminia Road, sometimes resulting in its closure while restoration occurs [63]. The publications also include engineering-geological data (e.g. rock unit weight, rock mass classification, joint friction angle) and rockfall runout tests which have been used to integrate and validate the results of this research.

3. Methodology

3.1. Topographic, TLS, and UAV Surveys

In order to recreate a detailed morphology of the slope, useful for the stability analysis and the rockfall runout modelling, an integrated TLS-UAV survey was carried out, allowing acquisition of a detailed 3D point cloud of the area. The integrated approach was fundamental given the complex morphology of the slope. This resulted in the limited use of TLS, restricting its deployment to the lower lateral parts of the rock mass, because of (i) the presence of water, which impedes data acquisition of the frontal parts of the slope, and (ii) the slope steepness, which impedes data acquisition in the higher parts of the slope. Therefore, the use of a UAV photogrammetric survey was essential to complete the data acquisition of the whole slope. On the other hand, the presence of vegetation and other obstacles precludes the possibility of a safe flight on the lateral lower parts of the rock mass. Therefore, the integration of UAV survey and TLS was essential to provide a complete model of the higher and the bottom zones of the slope and the interior of the tunnel. UAV photogrammetry was used extensively along the slope to map the inaccessible areas and provide overlapping data which increased the density and aided in the registration of the datasets.

TLS from 11 different positions was carried out by using a Trimble™ X8 device. After every scan, the scanner was alternated with a Nikon™ D7100 digital camera, properly mounted on a bracket (i.e., Leica™ Nodal Ninja 3II), to acquire high-definition (HD) images. Photos were shot using a fisheye lens (8 mm focal length); in this way, having a wide field of view (i.e., 180°), only six photos (every 60°) were necessary to guarantee an overlap of 66% and cover the whole panorama (the dip angle was set to −10°). In the laboratory, photos were processed using specific software for the creation of panoramic images 360° view (PTGui™—New House Internet Services BV). PTGui has the capability to model fisheye projections and to find reference points in the images. Having the focal center of the camera lens in correspondence with the optical center of the scanner, the HD photos were aligned (i.e., texture mapping) to the 3D point clouds of every scan position. Following this, a 3D photorealistic point cloud suitable for photointerpretation was created from the TLS survey. The different scans were relatively aligned using Iterative Closest Point (ICP) algorithm [65], while the absolute alignment was carried out by measuring the coordinates of nine pre-placed mobile optical targets located in strategic positions, visible from different TLS scans. All the georeferencing and texturing processes were carried out using Trimble™ RealWorks 10.4.

The adopted reference system was the Italian Gauss-Boaga, datum Monte Mario Italy 2. The coordinates of optical targets were measured by the combined use of TS (Leica™ Nova MS50) and two GNSS receivers (Leica™ GS15). The two GNSS devices firstly worked independently, in static modality, for a period longer than 3 h; then, the two measured points were processed using Leica™ Geo Office software and differential methods by combining simultaneous records from six other permanent GNSS stations of the Leica SmartNet ItalPos national network (i.e., R. San Marino/RSMARINO, Pesaro/PES2, Pennabili/PEN2, Gubbio/GUB2, Città di Castello/CIT1, Arcevia/ARCEVIA). The orthometric height of the measured points was calculated by using ConverGo software [66]. This procedure allowed a sub-centimetric accuracy for the two GNSS points, later necessary for georeferencing (i.e., roto-translation) the optical targets measured during the TS survey. In order to reach different parts of the slope, the TS was moved several times from its initial position by using the intersection method; this operation was carried out every time with an accuracy of about 1 mm.

The flights were performed using a DJI Phantom 4 PRO drone, equipped with a 20-megapixel digital camera with 1-inch sensor. The UAV surveys were carried out with direction of photo acquisition in nadiral modality (perpendicular to the lake) and in frontal modality (perpendicular and slightly oblique to the rock faces). This was done because SfM technology is based on sophisticated algorithms of image matching that use pseudo-random redundant images acquired from multiple viewpoints to reconstruct the three-dimensional geometry of an object or surface; hence, multiple images obtained from different angles help the image alignment procedure and limit non-linear deformations. Four flights were executed to cover all sectors of the slope, with UAV's distance from outcrops varied from 10 m up to 80 m, giving a nominal overlap and sidelap of at least 80% and 60%, respectively. An average estimated distance between pixel centers measured on the ground (i.e., Ground Sample Distance—GSD) of 1.8 cm was calculated.

The complex morphology of the area, with a deep and narrow gorge, did not allow for an automatic flight with waypoints: it was flown manually with an automatic triggering of photos every 2 s. Moreover, the use of a UAV equipped with a Real Time Kinematic (RTK) GNSS, which could have allowed acquisition of accurate coordinates of the camera location at every shot, was also not possible: the limited satellite visibility and the high Positional Dilution Of Precision (PDOP) within the gorge could have determined very low spatial accuracy in data. Therefore, the exterior orientation of all images (i.e., 1232 photos) was done by using 80 Ground Control Points (GCPs), which distribution is shown in Figure 2, measured again by the combining TS measurements, GNSS surveys in static modality, and differential data post-processing with a final millimeter-level positioning accuracy.

Eight artificial targets were pre-placed in the accessible zones of the study area, while the rest were obtained using natural features visible in the photos throughout the slope and by the TS during fieldwork. Their location was decided considering a balance between an optimum spatial distribution, both in space and elevation from the road, and easy identification of points on the images. The software Agisoft™ MetaShape (version 1.5.3; http://www.agisoft.com, last access: 24 June 2019) was used to process the images obtained from the UAV surveys. This software can solve the camera interior and exterior orientation parameters and generate georeferenced spatial data such as 3D point clouds, DEMs, DTMs and orthophotos. Agisoft™ MetaShape based on (i) SfM technology for the exterior orientation of images and the creation of the sparse 3D point cloud and (ii) MVS photogrammetry for the dense cloud reconstruction [67,68].

The first processing step consisted of image alignment, through which the interior and relative orientation parameters were solved. In order to improve the whole alignment process and to obtain low re-projection error, millions of tie points were automatically extracted without setting a point limit. Following image alignment, the second processing step involved georeferencing of the 3D model in such a way as to solve the exterior orientation parameters by using the GCPs coordinates measured during the GNSS-TS topographic surveys. The exterior orientation of images was necessary to measure the orientation, with respect to the North, and inclination of slopes and discontinuities which are needed for the stability analysis.

Figure 2. Perspective view, from SE, of Ground Control Point (GCP) spatial distribution along the slope (dense 3D textured point cloud in background).

Subsequently, the "optimize" tool of Agisoft™ MetaShape was utilized to adjust the estimated camera positions and to remove possible non-linear deformations, minimizing the errors due to re-projection and misalignment of the photos. Moreover, the optimization improved the model by deleting all the tie points with a re-projection error greater than one pixel.

In a subsequent step, the dense 3D point cloud was generated with medium quality and aggressive depth filtering settings. Automatic classification of cloud was done trying to remove vegetation, as much as possible, from the area of interest.

Finally, the different point clouds obtained by TLS and UAV image processing, both georeferenced in the Italian National Gauss-Boaga reference system, were unified using the open source CloudCompare™ software to produce a final complete 3D model of the rock mass including RGB data for texturing. The different point clouds were imported and no further alignment was executed. This is due to the fact the engineering-geological analyses were carried out on blocks and discontinuities only visible on the UAV-point cloud. The TLS-point cloud, instead, was fundamental for the complete characterization of the slope geometry, which is essential for the 3D rockfall runout analysis. As already mentioned, this was mainly due to the presence of trees on the lower part of the slope, which covered a portion of the rock mass in proximity to the local road. This caused distortions and holes on the UAV-point cloud, with possible consequences on the block trajectories. Therefore, the UAV-point cloud was replaced by the TLS-point cloud where needed (lower lateral parts of the slope). In this view, slight offsets between the two point clouds were considered negligible because, as explained later in Section 4.3.2, the spatial resolution of the input raster maps for the runout analysis was set equal to 50 cm/pixel, more than the possible error of mis-alignment of the two different point clouds.

3.2. Engineering-Geological Characterization of the Rock Mass and Stability Analysis

A direct engineering-geological survey was possible only on the lower parts of the slope. Unfortunately, the accessible areas allowed for the analysis of only few discontinuities, considered not fully representative of the conditions of the higher parts. For this reason, visual inspection in the field provided only a qualitative assessment of the rock mass conditions and the measurement of few joint characteristics (e.g., joint aperture, infill, weathering, compressive strength). Given the impossibility to collect all the necessary discontinuity properties, a proper characterization of the rock mass using traditional methods such as, for example, the Rock Mass Rating (RMR, [69,70]) was not possible. In the place of it, the Hoek and Brown Geological Strength Index (GSI, [71]) was determined since it bases on qualitative geological observations.

The structural analysis of the higher parts of the slope was carried out by using the Compass plugin [72] of the freeware CloudCompare™ software. Compass is a plugin which allows measurement of discontinuity orientations directly on the point cloud. In this way, the orientation of about 70 discontinuities in strategic areas of the slope was acquired to create a stereonet used to identify the different discontinuity sets. Following this, a kinematic stability analysis, based on the Markland test [73] was performed. Since the results of the kinematic analysis are influenced by geotechnical properties and topographic factors (i.e., dip and dip direction of rock outcrops), the slope was divided in four sectors (Figure 3) with different orientations.

Figure 3. Different slope directions identified and used for the kinematic stability analysis; orthophoto from thee UAV survey in the background.

Such analysis allowed identification of the main types of failure (i.e., toppling, planar sliding, wedge sliding).

In addition, CloudCompare was used to estimate volumes, geometries, and positions of all potentially unstable blocks that could trigger rockfall events in the future. This provided a basis to identify source blocks to be used in the subsequent dynamic stability analysis and rockfall runout simulation. The geometric information can be extracted from the point cloud manually, or utilizing software to automate the process [74,75]. For example, in [76] the authors used Polyworks (InnovMetric Software, 2017) to define the sets of the discontinuities based on 3D point clouds and then used the spacing of each set to roughly calculate the average size of blocks. In their work, the discontinuity information was extracted manually from the point clouds and the block size was calculated approximately. Differently, in [77] a fully automatic method for extracting rock block information from 3D point clouds was presented.

In this paper blocks were identified on the final point cloud by a careful photointerpretation that was also revised and validated by experts. The following procedure was used: (i) the point cloud was segmented in several areas containing a potentially unstable block each; (ii) every joint located below a potentially sliding block was sampled and co-planar points selected; (iii) the cloud representing each block was cut and the normals to the points, oriented toward the exterior, computed; (iv) the clouds representing both the block and the underlying joint were merged and used to build a mesh through the Poisson Surface Reconstruction Plugin [78]; (v) finally, by means of the editing tools of CloudCompare, the volume of every mesh was measured.

The dynamic stability analysis of rock blocks was carried out by Rocscience™ codes (i.e., Rocplane, Swedge, Roctopple for planar sliding, edge sliding and block toppling, respectively). SFs were calculated both in static and dynamic conditions considering the presence of water, and the local seismic acceleration. Using GeoStru™ PS software (GeoStru, 2019) the maximum horizontal acceleration "ag" of the site was derived (i.e., 0.25 g in this study). The procedure was carried out following the new Italian "Norme Tecniche per le Costruzioni" [79].

3.3. Rockfall Runout Analysis

The mathematical model that can be adopted to analyze the rockfall runout must be able to describe the behavior of a falling block in terms of trajectory direction and length, bounce height, speed, energy, and type of movement. Two principal methods can be used to solve this problem: (i) lumped mass methods and (ii) rigorous methods [80]. The lumped mass approach considers the single block as a point with specific mass and velocity. Therefore, it is a simplified model where the velocity of the block is reduced after every impact with the surface by using two coefficients Kn (normal coefficient of restitution) and Kt (tangential coefficient of restitution). Differently, a more rigorous approach considers the shape and dimension of the falling block, which must be known a priori. It is a complete method that considers the moment changes, depending on shape of block and surface, angle of trajectory, roughness of the surface, and friction between block and surface [80].

The method used in this paper follows a 3D rigorous approach. The simulation software is called Rockyfor3D [81] and combines physically based deterministic algorithms with stochastic approaches. In this sense, Rockyfor3D is also defined as "probabilistic process-based rockfall trajectory model." The trajectory is simulated as a 3D vector data by sequences of parabolic free fall through the air and rebounds on the surface. Sliding of the rock is not considered, as well as block fragmentation, and rolling of the mass is represented by a sequence of short-distance rebounds [81]. Moreover, the model considers impacts against trees. The simulations can be run by using 14 ASCII rasters characterized by the same cell size and extent, which describes the characteristics of slope surface and falling blocks. The input raster maps, created in a GIS environment using the data from the geomatic and geological surveys, are described in Table 1.

The runout simulations with the current slope configuration were performed using the input blocks identified on the UAV images and measured on the 3D point cloud. However, it has to be mentioned that, before proceeding with the rockfall runout simulation, a model calibration step was done; this process, also defined as "back analysis" in respect of a past rockfall event, which is essential to calibrate the entire 3D model together with the simulation input parameters and raster. In this case, given the steep geometry of the slope, which is characterized essentially by steep bedrock with limited vegetation overhanging water dump, it is difficult to simulate a past rockfall event, since the arrival point is generally missing (i.e., below water level). In terms of rockfall runout analysis, well-represented geometrical info were obtained: not only slopes, but also the potentially unstable blocks were well known in terms of shape and size. Physical parameters were calibrated (i.e., coefficients of normal and tangential restitution) in such a way that input data can be considered adequate for a reliable rockfall runout study.

Table 1. Input raster maps necessary for using Rockyfor3D.

Input Raster	Description
DTM	It represents the DTM of the slope.
Rock density	It represents the rock density of every simulated source cell (i.e., falling block).
Block height, width and length	These three raster maps describe the shape of a block, in terms of height, width, and length.
Block shape	It defines the shape of a falling block (i.e., rectangular, ellipsoidal, spherical, and disc shaped).
Roughness 70, 20, 10	These three raster maps define the roughness of the slope surface, represented by all the obstacles lying on the slope. The value of every cell, expressed in meters, corresponds to the height of a representative obstacle that can be encountered by a block with a probability of 70% (rg70), 20% (rg20) and 10% (rg10). The values range from 0 to 100 and are used to calculate the tangential coefficient of restitution (Rt) causing an energy loss.
Soil type	This raster map defines the elasticity of the ground surface basing on the soil type; there are eight different soil types (e.g., bedrock, asphalt road, fine soil material) and every type corresponds to a certain value of the normal coefficient of restitution (Rn).

4. Results

4.1. D Modelling

The final integrated point cloud (TLS + UAV) is composed of 22,308,471 points, corresponding to an average point spacing of 1.91 cm. The 3D model was built in mesh format constituting of 18,753,142 facets. Even if the final obtained 3D model is considered suitable for the goals of the study, it must be mentioned that the two point clouds come from different processing workflows (TLS and SfM-MVS photogrammetry), using different control points used for georeferencing and, therefore, have diverse spatial accuracy. The point cloud from TLS has a Registration Error of 5.45 mm (i.e., mis-alignment among the 11 scans) and a Georeferencing Error of 10 mm in respect to the artificial targets. The point cloud from SfM-MVS photogrammetry has a Root Mean Square Error (RMSE) of 80 mm in respect of GCPs. It is assumed that this higher error is due to the lack of targets properly fixed along the slope and the uncertainty in defining them on the UAV images.

4.2. Stability Analysis

4.2.1. Engineering-Geological Characterization

As already mentioned in Section 3.2, the traditional engineering-geological survey was limited due to inaccessibility of the study areas. Therefore, a combined approach was used with the interpretation of the point cloud from remote sensing data. The rock mass conditions, estimated through the GSI by field inspections and interpretation of photographs from UAV, show a difference between the bottom and the higher parts of the slopes. In this regard, GSI is considered to be varying from 70–80 (lower parts) to 50–60 (mid-higher parts). In fact, it was possible to observe how the rock slope is influenced (especially in the mid and higher parts) by several fractures related to different joint systems. A total of 67 discontinuities were measured, in terms of dip and dip direction, directly on the final point cloud through the CloudCompare™ Compass plugin. Then, the presence of joint sets was determined by means of a density analysis of attitudes in stereographic projection: the lower hemisphere of the Schmidt (equal area) net was used (Figure 4). Table 2 summarizes the orientation values of each joint set.

Results from the engineering-geological analysis, revealed a certain variability of the discontinuity pattern, with a difference between the basal and the mid-higher parts of the rock slope. In addition, the discontinuities dip angles increase, sometimes reversing their direction (an example is shown in Figure 5). Figure 5 shows the identification of the main discontinuities on the mid parts of the slope, the area most subjected to unstable block formation.

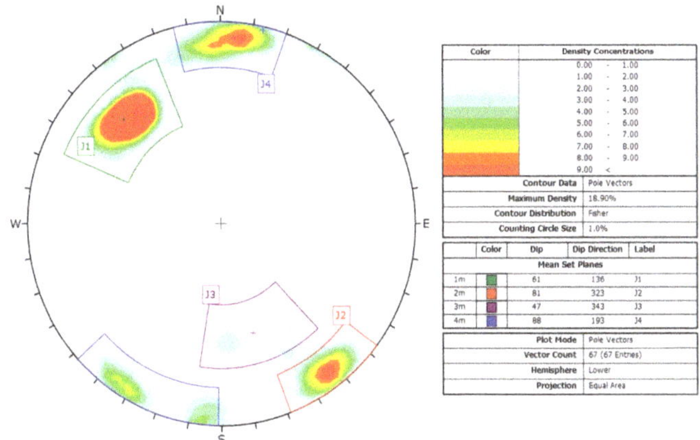

Figure 4. Final stereonet plot of discontinuity poles measured from remote sensing and traditional surveys: joint set sub-division after density analysis is highlighted.

Figure 5. Examples of the main discontinuity sets identified on the mid-part of the rock slope.

Table 2. Mean orientation values of discontinuity sets measured on the study area.

Set	Mean Dip	Mean Dip Direction
J1	61	136
J2	81	323
J3	47	343
J4	88	193

4.2.2. Kinematic Stability Analysis

After identification of the main discontinuity sets, a kinematic stability analysis was carried out to identify the possible failure conditions on the slope. This type of analysis is based on discontinuity orientations, slope direction and discontinuity friction angle. As previously explained, the discontinuity orientation and slope direction were derived from the point cloud, while the friction angle of the failure surfaces (25°) was chosen based on previous studies on the same area [63], the authors expertise, and conservative geotechnical assumptions. Table 3 summarizes the results of the kinematic analysis for the identified four main slope directions.

Results shown in Table 3 indicate that along V1 the formation of planar sliding (on J1) and wedge sliding (on the intersection between J1–J4) are possible (example in Figure 6a). With the same failure modes present in V2, with the addition of possible direct toppling (example in Figure 6b). Differently, on V3 only planar sliding on J1 is possible. Along V4 wedge sliding due to the intersection between J3 and J4 is possible.

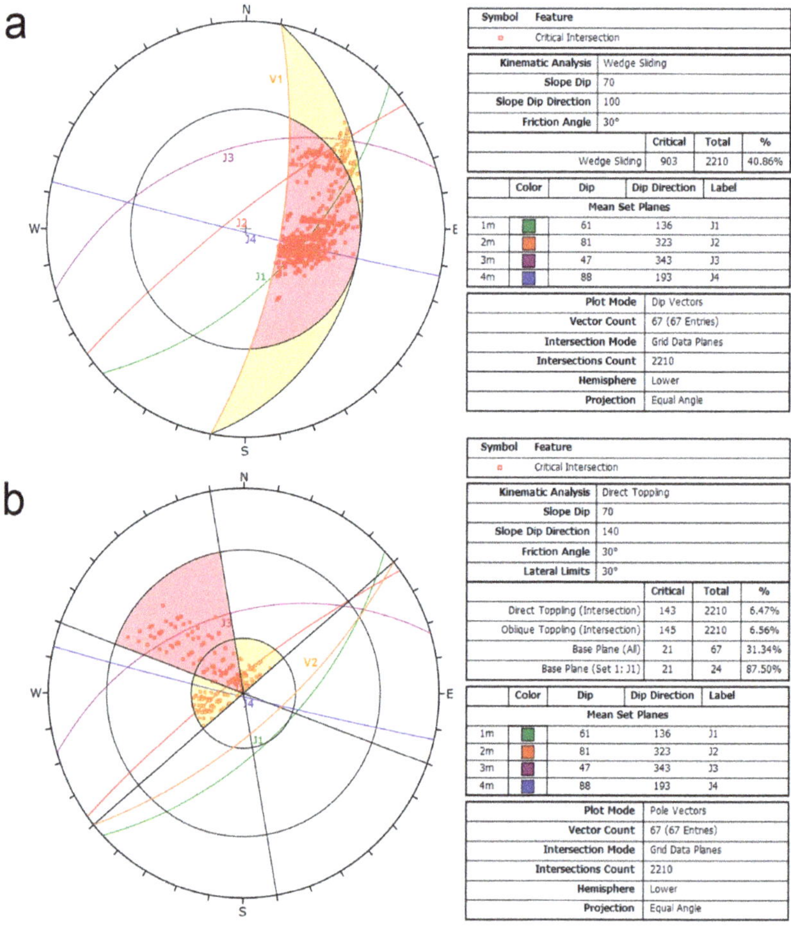

Stereonet **a**:

Symbol	Feature
□	Critical Intersection

Kinematic Analysis	Wedge Sliding
Slope Dip	70
Slope Dip Direction	100
Friction Angle	30°

	Critical	Total	%
Wedge Sliding	903	2210	40.86%

	Color	Dip	Dip Direction	Label
		Mean Set Planes		
1m		61	136	J1
2m		81	323	J2
3m		47	343	J3
4m		88	193	J4

Plot Mode	Dip Vectors
Vector Count	67 (67 Entries)
Intersection Mode	Grid Data Planes
Intersections Count	2210
Hemisphere	Lower
Projection	Equal Angle

Stereonet **b**:

Symbol	Feature
□	Critical Intersection

Kinematic Analysis	Direct Toppling
Slope Dip	70
Slope Dip Direction	140
Friction Angle	30°
Lateral Limits	30°

	Critical	Total	%
Direct Toppling (Intersection)	143	2210	6.47%
Oblique Toppling (Intersection)	145	2210	6.56%
Base Plane (All)	21	67	31.34%
Base Plane (Set 1: J1)	21	24	87.50%

	Color	Dip	Dip Direction	Label
		Mean Set Planes		
1m		61	136	J1
2m		81	323	J2
3m		47	343	J3
4m		88	193	J4

Plot Mode	Pole Vectors
Vector Count	67 (67 Entries)
Intersection Mode	Grid Data Planes
Intersections Count	2210
Hemisphere	Lower
Projection	Equal Angle

Figure 6. Examples of kinematic analysis for wedge sliding along V1 (**a**) and direct toppling along V2 (**b**).

Table 3. Results of the kinematic stability analysis along four identified slopes.

Versant	Planar Sliding	Wedge Sliding	Direct Toppling
	Critical System or Critical Intersection		
V1 (100/70)	J1	J1–J4	/
V2 (140/70)	J1	J1–J4	J2–J4–J1
V3 (180/70)	J1	/	/
V4 (220/70)	/	J3–J4	/

4.3. Rockfall Analysis

4.3.1. Unstable Block Identification and Dynamic Stability Analysis

The kinematic analysis provided an understanding of possible block formations useful for the geological hazard assessment of the area. Attention was placed on the dynamic stability conditions and the rockfall runout analysis of seven potentially unstable blocks/rock masses (A, B, C, D, F, E, K in Figure 7) that were identified on the textured point cloud. Their geometrical characteristics are summarized in Table 4 together with the most critical SF as calculated in dynamic conditions considering the presence of water and seismic forces.

Figure 7. Identification of critical blocks on the slope.

Table 4. Geometrical characteristics of unstable blocks identified on the slope and their most critical Safety Factors (SFs). * = SF not identified due to the presence of past remediation works (i.e., covering network and artificial reinforcement).

ID	Volume (M³)	Discontinuity		Discontinuity Set				Block Centroid Coordinates (M)			Most Critical SF (Dynamic Condition)
		Dip Direction (°)	Dip (°)	J1	J2	J3	J4	X	Y	Z	
A	12.41	129 005	58 80	X			X	2,337,011.05	4,835,874.36	223.76	1.03
B	0.25	108	63	X				2,337,018.14	4,835,881.80	221.24	*
C	2.88	129 181	58 75	X			X	2,337,014.10	4,835,876.41	220.97	1.20
D	11.50	202	59					2,337,006.44	4,835,873.76	225.89	1.08
E	13.14	108	63	X				2,337,014.07	4,835,881.93	229.17	1.09
F	5.92	129 006	58 80	X			X	2,337,012.77	4,835,875.91	223.69	1.10
K	2.39	175 140	81 50			X	X	2,337,006.54	4,835,871.14	205.26	1.01

4.3.2. Rockfall Runout and Hazard Assessment

As already mentioned in the Methods section, before proceeding with the rockfall runout simulation, a back analysis in respect of a past rockfall event was executed to properly calibrate the 3D model, together with the simulation input parameters and raster. This analysis was done considering reliable trajectories in which end-points were located along the Via Flaminia road, in the archaeological area, and within the lake. After this, the simulations were performed by considering each identified block, assuming real block dimensions measured directly on the point cloud. The final simulations were configured as follows: (i) number of trajectories for each block equal to 50; (ii) block volume weight 2500 kg/m³; (iii) random variation of the block volume equal to ±10%. The spatial resolution of the input raster maps was set equal to 50 cm/pixel. Experiences reported in [82] showed that the preferred resolution lies between 2 × 2 m and 10 × 10 m. In addition, a 1 × 1 m resolution does not necessarily improve the quality and it increases the amount of data substantially. In respect to this, the cell size adopted in this work is already considered too detailed.

Some representative outputs of the rockfall runout simulation are shown in Figure 8. The figure shows the results in terms of:

- Kinetic energy at the 95% confident level (translational + rotational, in kJ) of all blocks that passed over a single cell.
- Passing height values at the 95% confident level (measured from the block barycenter in normal direction to the slope surface, in meters).
- Probability for a block to reach a specific cell (i.e., (Nr. Passages × 100)/(Nr. Simulations per source cell × Nr. Source cells)).

In order to evaluate the rockfall geological hazard for the area, the values of probability, kinetic energy, and passing heights of each block run in the simulations were re-classified as shown in Table 5. These values were obtained based on expertise and knowledge of characteristics of the net barriers already present in the area. The values of energy, for example, were chosen considering the potential to stop blocks within a range between 1100 kJ (as minimum) and 5000 kJ (as maximum), representative of the capacity of a standard barrier available on the market.

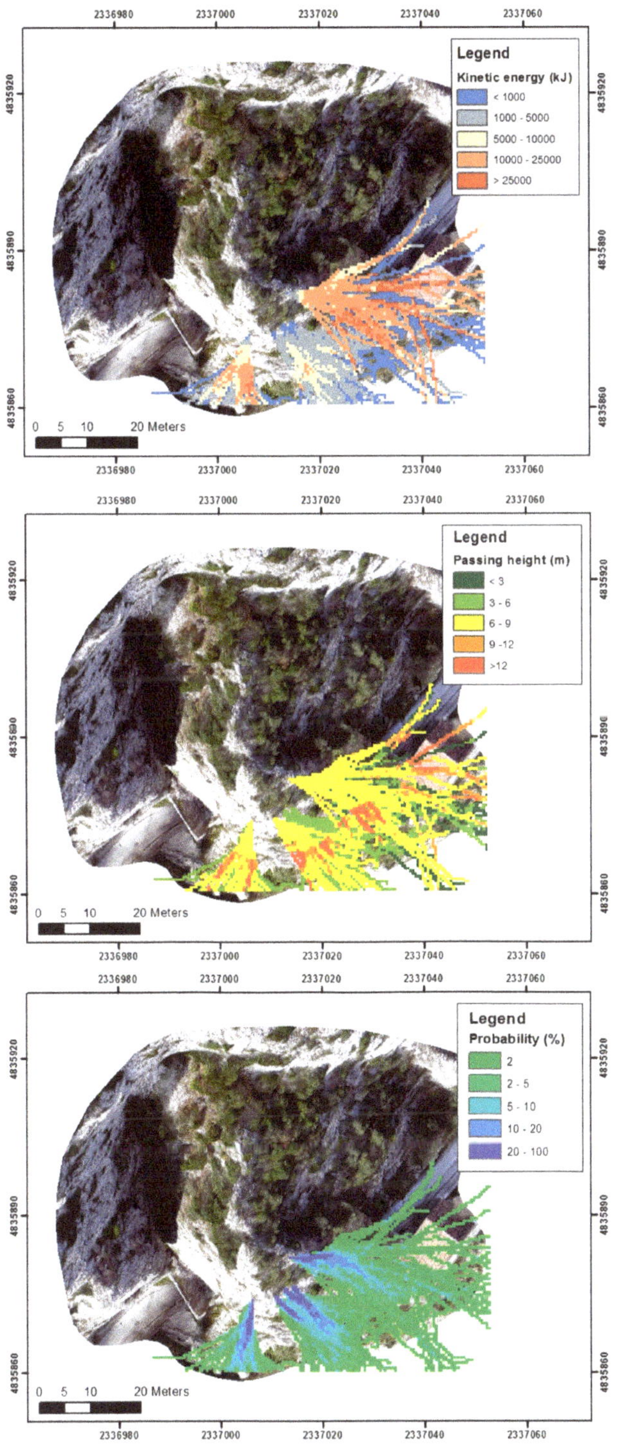

Figure 8. Kinetic energy, passing height, and probability maps of the identified falling rock blocks.

Table 5. Classification of energy, height, and probability raster outputs.

Energy (kJ)		Height (m)		Probability (%)	
Class	Value	Class	Value	Class	Value
1	≤1100	1	≤2	1	≤10
2	1100 < kJ ≤ 5000	2	2 < m ≤ 5	2	10 < % ≤ 20
3	>5000	3	>5	3	20 < % ≤ 100

By the combination of the re-classified raster maps, three different classes of hazard were established (i.e., low, medium, high). A preliminary hazard class was assigned to each cell based on the combination of Height and Energy (Figure 9). Subsequently, this preliminary class was combined with the Probability value in order to define the final rockfall hazard class. For example, a block having high energy but low rebound height and low probability of reaching a certain point of the slope would be assigned to the medium hazard class. Differently, a block with medium rebound height and energy and a high reach probability, would be assigned to the high hazard class. The final hazard map obtained from the described approach is shown in Figure 10.

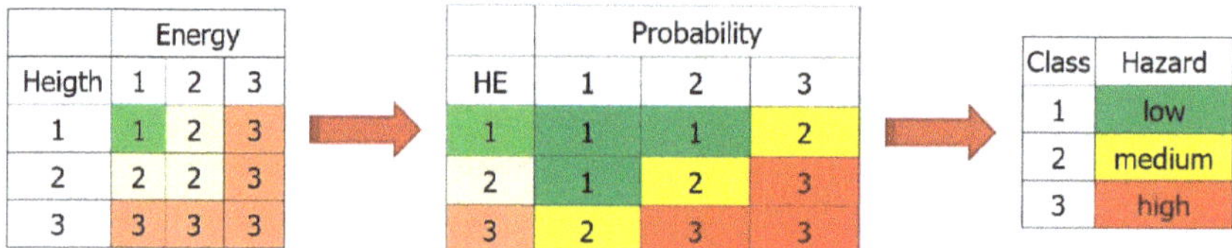

Figure 9. Criteria of classification used to create the rockfall hazard map.

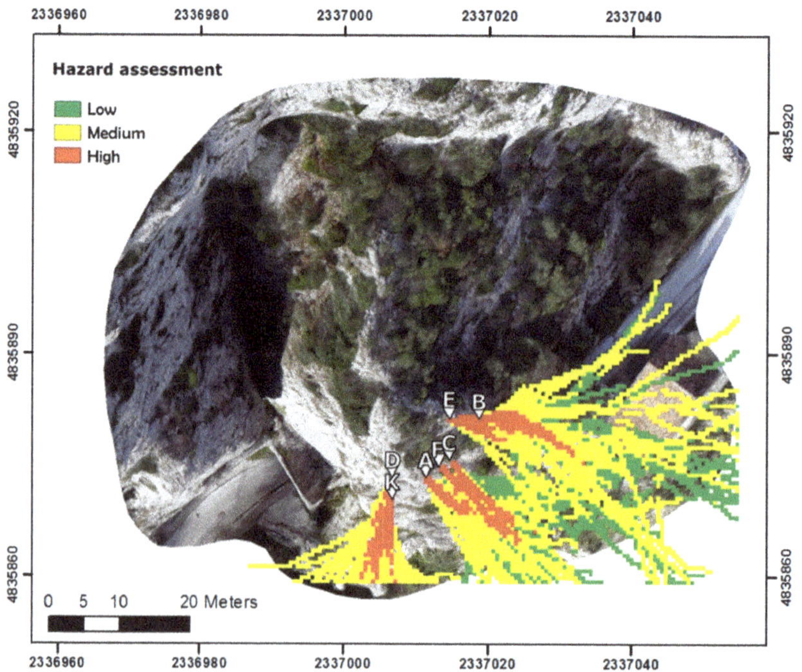

Figure 10. Rockfall hazard map.

4.3.3. Simulation with Virtual Net Counters

With the aim of proposing rockfall control defenses for the ancient Roman Flaminia road and the church, an additional simulation with virtual net counters was run. A greater number of simulations were performed (i.e., 1000 trajectories/source point) with the aim of increasing the reliability of the analysis [83]. Four net counters were included in the simulations as shown in Figure 11. The nets were virtually placed at the bottom of the rock slopes, collecting detailed data relative to (i) number of overpassing blocks, (ii) kinetic energy (95th percentile of the simulated energy), (iii) passing heights (95th percentile of the simulated passing heights), and (iv) reach probability for each net.

Figure 11. Scheme of virtual net counters.

The analysis indicated that it is theoretically possible to define the most suitable position where to localize rockfall barriers or other remediation measures. Results of this simulation are summarized in Table 6.

Table 6. Results of rockfall simulation with virtual net counters.

Block ID	Net ID	Energy (kJ)	Height (m)	Number of Passages	Probability (%)
A	2	6394.9	11.7	817	81.7
	3	191.1	2.3	1	0.1
B	2	132.7	2.7	25	2.5
	3	192.9	5.9	850	85
	4	221.5	7.1	76	7.6
C	2	2595.1	9.1	987	98.7
	3	1113.7	1.7	3	0.3

Table 6. *Cont.*

Block ID	Net ID	Energy (kJ)	Height (m)	Number of Passages	Probability (%)
D	1	9174	11.5	953	95.3
	2	5054	3.6	3	0.3
E	2	22,229.3	6.9	41	4.1
	3	27,054.5	8.4	553	55.3
	4	25,881.5	9.8	43	4.3
F	2	1458	5.8	987	98.7
	3	1162.3	1.7	3	0.3
K	1	1005	5.8	997	99.7

5. Discussion

The study area is a steep slope characterized by the presence of four different discontinuity sets, that isolate the portion of rock mass forming blocks prone to collapse. Considering that a large part of the slope is inaccessible, the use of geomatic techniques was essential in this study for geological mapping purposes. Indeed, it was possible to remotely collect data about blocks and discontinuities to be used for a detailed geometric characterization of the slope in a spatially uniform way throughout the area. However, it should be stated that without a classical engineering-geological survey, fundamental properties of discontinuities and rock would be missing (e.g., joint aperture, infill, weathering, roughness). This is an aspect where future research could concentrate in developing new technologies and approaches, but the contribution of traditional surveys is today still essential for a correct characterization of a rock mass.

The integration of data from UAV photogrammetry and TLS was applied in this study with the aim of producing a detailed and complete 3D model of the slope, used for collecting accurate geometrical information that is needed for the stability analysis and rockfall runout simulation. To achieve this, the TLS was performed since it is a rapid technique that ensures at the same time high spatial resolution and accuracy of measured data. UAV photogrammetry was utilized because it allows the interpretation of the higher parts of the slope which are inaccessible and not visible from the Via Flaminia road. Moreover, the use of UAV was essential for the study purposes in consideration of the narrow shape of the gorge, the total inaccessibility of surrounding slopes and the lack of an open space from which a complete survey from higher position was achieved. In addition, photogrammetry provided a complete inspection and photointerpretation in a cost-effective manner, in just few hours of fieldwork. The final point cloud, representative of the whole slope including the ancient Roman Via Flaminia tunnel and the "Santa Maria delle Grazie" Church with a centimetric level of accuracy, allowed acquisition of deterministic discontinuity information (dip and dip direction). This helped in identifying the discontinuity sets (Figure 4), from which the kinematic stability analyses were derived. In particular, the slope is characterized by a minor presence of discontinuities in the lower parts, while most discontinuities are present in the mid-higher parts, where they show also higher dip slope, with few examples of local overturning (Figure 5).

Interpretation of results from the kinematic analysis (Table 3) revealed also that the most probable failure mode is by planar sliding on J1 and wedge sliding along the intersection between J1 and J4 sets.

The final point cloud was also used to identify unstable blocks and measure their geometry. The dynamic stability analysis revealed that rock blocks A, D, E, and K are at a limit of equilibrium (SF $\cong$ 1.0), while the area of block B, that overheads the Church, needs cleaning and stripping of small stone elements that are present. It must be noted that the present study refers to current conditions which are unpredictably changeable (e.g., water, ice, chemical-physical alteration, interactions between different blocks). At present, the blocks and wedges identified in the area of interest clearly show the presence of rock bridges that favor the stability.

The runout analysis of the seven blocks identified thanks to the geomatic data, indicate the possibility that falling blocks could reach the local road, the archaeological site and the adjacent Church. In detail, results show low probability for the local road to be reached by a block (<4.5% for block E and <8% for block B taking as reference the virtual net counter number 4). On the contrary, the adjacent Church shows a higher theoretical probability (i.e., 85% for block E and 55% for block B taking as reference the virtual net counter number 3). A significant risk is associated to the reference virtual net counters 1 and 2 with possible rockfall of blocks A, C, D, F, and K (probability generally higher than 80% or 90%). However, these values do not consider the return period parameter and the probability that a single block could fall. Nevertheless, this can be considered as a conservative approach, given that if those parameters had been considered in the calculation, the probability would have certainly decreased. Moreover, it must be mentioned that the used code for rockfall runout simulation (i.e., Rockyfor3D) does not consider block fragmentation that could influence falling element trajectory, velocity, energy, bounce height, and end point. This is in agreement with the conservative approach of the study, and it could be included in future analyses.

The information collected during this study will be used by local authorities for developing projects of remediation works. Regarding this, the virtual net counters provide knowledge of the energy of every block as well as the passing height and velocity, which are fundamental information when planning rockfall protection measures. Results of the rockfall simulation showed how some of the falling blocks can produce high energies, high rebounds, or both. For example, the analysis showed how blocks E and D may reach the local road or the adjacent Church with a high energy, superior to the capability of a standard net barrier available on the market. On the other hand, blocks A and C showed lower energy, but high bounce heights that could pass an eventual barrier. Consequently, a possible immediate measure could be the removal or the consolidation by means of a harness with ropes, or with panels of cables laterally anchored on the stable rock, of blocks A, C, D, and E. Blocks smaller than 3 m^3 could be destroyed or removed.

More generally, the following actions could be carried out:

- inspection, surface cleaning, and controlled barring for blocks smaller than 1 m^3;
- destruction and removal of the smallest blocks;
- wire mesh with reinforced ropes;
- wire mesh panels;
- rock bolts, harness with ropes, or wire rope panels;
- sealing/consolidation with suspensions and consolidating mixtures;
- reinforcing wall construction;

Results from this study must be used to focus further analyses on the segments with higher probability to be reached by a block, for a proper design of protecting net barriers. This is very important in view of the fact that the hazard map obtained from the analysis (Figure 10) shows a medium to high hazard for the entire area. This is critical as the area is a popular place given the presence of the archaeological site, the Church, and a local ancient Roman road

6. Conclusions

This case study has shown an application of a combined working approach that integrates the use of traditional engineering-geological techniques and geomatic technologies with the aim of carrying out a 3D rockfall runout analysis of a natural slope in Italy. The study confirms how integration of UAV photogrammetry and TLS data can be useful to overcome the common problems found in traditional approaches, such as the lack of data of adequate scale factor, the site accessibility, the unsafe conditions, and the complete identification of possible rockfall source points. TLS or UAV photogrammetry, alone, would not have produced a complete model of this area: only their integration, carried out respecting the basic guidelines of data processing and georeferencing, could guarantee the full and reliable modelling of the slopes.

Several discontinuities and rock blocks were identified above the archaeological area and near the Church of "Santa Maria delle Grazie" along the ancient Roman Via Flaminia road. The complete 3D model of the slope provided a means to characterize the rock mass discontinuity patterns and define the possible falling trajectories of each block. At present, each wedge is characterized by theoretical limit equilibrium conditions (with possible presence of unknown rock bridges), so that the adoption of some remediation measures (e.g., barring, chemical desegregation, removal, net barriers installation) is pointed out. In the opinion of the authors, the main advantages of the adopted approach for rockfall runout analysis, even in consideration of the study aims, can be summarized as it follows:

- Provides the possibility to obtain deterministic geometrical data about rock blocks (i.e., position on the slope, shape and volume) and discontinuities (i.e., dip and dip direction) in inaccessible areas with short working times and in a cost-effective manner thanks to SfM-MVS photogrammetry and UAV surveys.
- Provides the possibility to assess the stability conditions of single blocks, to compute their SFs, and to identify possible rockfall sources, thanks to the acquisition of detailed information about rock wedges volume and weight, slope orientation, and rock mass conditions.
- Provides the possibility to perform geological hazard studies that identify areas with high probability to be reached by rockfall events.
- Provides the possibility to simulate the presence of net counters that allow collection of information regarding energy, passing heights, and reach probability of blocks that can be used as quantitative data for protection measures planning.

However, it must be said that some limitations are also present in such a study. One main issue with SfM-MVS photogrammetry is quantifying the uncertainty and accuracy without the use of a number of GCPs. Especially in zones where the narrow shape of the gorge makes impossible a proper use of UAV equipped with an RTK-GNSS, the number of GCPs should be higher than usual (such as this case study). This, combined with the height of slopes, makes the work more difficult despite an extensive and extreme use of TS. Laser scanning from a UAV represents an important tool and an opportunity in case studies as the one presented in this paper since it can acquire 3D point clouds with constant spacing and high density. Nevertheless, the use of images, and particularly the adoption of SfM-MVS photogrammetry, gives the advantage of making possible the photointerpretation and the deep study of features.

The presented case study highlights the contribution of geomatics in the interpretation and analysis of engineering-geological investigations and demonstrates a way to overcome practical issues of traditional survey techniques related to the need of accurate and detailed data and the difficult access to rock slopes.

Author Contributions: Investigation, Claudio Vanneschi, Filippo Bonciani, Marco Di Camillo, and Riccardo Salvini; Formal Analysis, Eros Aiello, Filippo Bonciani, and Riccardo Salvini; Data Curation, Claudio Vanneschi; Methodology and Validation; Claudio Vanneschi, and Riccardo Salvini; Supervision Eros Aiello, and Riccardo Salvini; Writing—original draft, Claudio Vanneschi, Marco Di Camillo, and Riccardo Salvini; Writing—review & editing, Claudio Vanneschi, and Riccardo Salvini.

Funding: This research received no external funding.

Acknowledgments: Authors appreciatively acknowledge the grant of this project from Enel Green Power O&M Hydro Italy—Southern Central Area—and the assistance from professionals with hydro-civil function of the Furlo dam. A special thank goes to our colleagues Barbarino M., Cinà A. and Di Bartolo S. for their contribution to the research and the assistance during fieldwork.

Conflicts of Interest: The authors declare no conflict of interest.

References

1. Hungr, O.; Evans, S.G.; Hazzard, J. Magnitude and frequency of rock falls and rock slides along the main transportation corridors of southwestern British Columbia. *Can. Geotech. J.* **1999**, *36*, 224–238. [CrossRef]

2.	Gigli, G.; Morelli, S.; Fornera, S.; Casagli, N. Terrestrial laser scanner and geomechanical surveys for the rapid evaluation of rock fall susceptibility scenarios. *Landslides* **2014**, *11*, 1–14. [CrossRef]

3.	Lim, C.H.; Martin, C.D.; Herd, E.P.K. Rock fall hazard assessment along railways using GIS. In Proceedings of the Canadian Geotech. Conference, Québec City, QC, Canada, 24–27 October 2004; pp. 1–8.

4.	Lan, H.; Martin, C.D.; Zhou, C.; Lim, C.H. Rockfall hazard analysis using LiDAR and spatial modeling. *Geomorphology* **2010**, *118*, 213–223. [CrossRef]

5.	Pfeiffer, T.J.; Bowen, T. Computer simulation of rockfalls. *Bull. Assoc. Eng. Geol.* **1989**, *26*, 135–146. [CrossRef]

6.	Azzoni, A.; La Barbera, G.; Zaninetti, A. Analysis and prediction of rockfalls using a mathematical model. *Int. J. Rock Mech. Min. Sci. Geomech. Abstr.* **1995**, *32*, 709–724. [CrossRef]

7.	Antoniou, A.A.; Lekkas, E. Rockfall susceptibility map for Athinios port, Santorini Island, Greece. *Geomorphology* **2010**, *118*, 152–166. [CrossRef]

8.	Youssef, A.M.; Pradhan, B.; Al-Kathery, M.; Bathrellos, G.D.; Skilodimou, H.D. Assessment of rockfall hazard at Al-Noor Mountain, Makkah city (Saudi Arabia) using spatio-temporal remote sensing data and field investigation. *J. Afr. Earth Sci.* **2015**, *101*, 309–321. [CrossRef]

9.	Descoeudres, F. Three-dimensional dynamic calculation of rockfalls. In Proceedings of the International Congresson on Rock Mechanics, Montreal, QC, Canada, 30 August–3 September 1987.

10.	Agliardi, F.; Crosta, G. High resolution three-dimensional numerical modelling of rockfalls. *Int. J. Rock Mech. Min. Sci.* **2003**, *40*, 455–471. [CrossRef]

11.	Lan, H.; Martin, C.D.; Lim, C. RockFall analyst: A GIS extension for three-dimensional and spatially distributed rockfall hazard modeling. *Comput. Geosci.* **2007**, *33*, 262–279. [CrossRef]

12.	Chiessi, V.; D'Orefice, M.; Mugnozza, G.S.; Vitale, V.; Cannese, C. Geological, geomechanical and geostatistical assessment of rockfall hazard in San Quirico Village (Abruzzo, Italy). *Geomorphology* **2010**, *119*, 147–161. [CrossRef]

13.	Thoeni, K.; Giacomini, A.; Lambert, C.; Sloan, S.W.; Carter, J.P. A 3D discrete element modelling approach for rockfall analysis with drapery systems. *Int. J. Rock Mech. Min. Sci.* **2014**, *68*, 107–119. [CrossRef]

14.	Fanos, A.M.; Pradhan, B. A novel rockfall hazard assessment using laser scanning data and 3D modelling in GIS. *Catena* **2019**, *172*, 435–450. [CrossRef]

15.	Moos, C.; Fehlmann, M.; Trappmann, D.; Stoffel, M.; Dorren, L. Integrating the mitigating effect of forests into quantitative rockfall risk analysis–Two case studies in Switzerland. *Int. J. Disaster Risk Reduct.* **2018**, *32*, 55–74. [CrossRef]

16.	Fanti, R.; Gigli, G.; Lombardi, L.; Tapete, D.; Canuti, P. Terrestrial laser scanning for rockfall stability analysis in the cultural heritage site of Pitigliano (Italy). *Landslides* **2013**, *10*, 409–420. [CrossRef]

17.	Farvacque, M.; Lopez-Saez, J.; Corona, C.; Toe, D.; Bourrier, F.; Eckert, N. How is rockfall risk impacted by land-use and land-cover changes? Insights from the French Alps. *Glob. Planet. Chang.* **2019**, *174*, 138–152. [CrossRef]

18.	Shen, W.-G.; Zhao, T.; Crosta, G.B.; Dai, F. Analysis of impact-induced rock fragmentation using a discrete element approach. *Int. J. Rock Mech. Min. Sci.* **2017**, *98*, 33–38. [CrossRef]

19.	Lopez-Saez, J.; Corona, C.; Eckert, N.; Stoffel, M.; Bourrier, F.; Berger, F. Impacts of land-use and land-cover changes on rockfall propagation: Insights from the Grenoble conurbation. *Sci. Total. Environ.* **2016**, *547*, 345–355. [CrossRef] [PubMed]

20.	Hutchinson, J.N. Morphological and geotechnical parameters of landslides in relation to geology and hydrogeology, state-of-the-art report. In Proceedings of the 5th International Symposium on Landslides, Lausanne, Switzerland, 10–15 July 1988; Bonnard, C., Ed.; pp. 3–35.

21.	Evans, S.; Hungr, O. The assessment of rockfall hazard at the base of talus slopes. *Can. Geotech. J.* **1993**, *30*, 620–636. [CrossRef]

22.	Dorren, L.K. A review of rockfall mechanics and modelling approaches. *Prog. Phys. Geogr. Earth Environ.* **2003**, *27*, 69–87. [CrossRef]

23.	Giani, G.P.; Giacomini, A.; Migliazza, M.; Segalini, A. Experimental and Theoretical Studies to Improve Rock Fall Analysis and Protection Work Design. *Rock Mech. Rock Eng.* **2004**, *37*, 369–389. [CrossRef]

24.	Bellian, J.; Kerans, C.; Jennette, D. Digital Outcrop Models: Applications of Terrestrial Scanning Lidar Technology in Stratigraphic Modeling. *J. Sediment. Res.* **2005**, *75*, 166–176. [CrossRef]

25. Lague, D.; Brodu, N.; Leroux, J. Accurate 3D comparison of complex topography with terrestrial laser scanner: Application to the Rangitikei canyon (N-Z). *ISPRS J. Photogramm. Remote Sens.* **2013**, *82*, 10–26. [CrossRef]

26. Brothelande, E.; Lénat, J.-F.; Normier, A.; Bacri, C.; Peltier, A.; Paris, R.; Kelfoun, K.; Merle, O.; Finizola, A.; Garaebiti, E. Insights into the evolution of the Yenkahe resurgent dome (Siwi caldera, Tanna Island, Vanuatu) inferred from aerial high-resolution photogrammetry. *J. Volcanol. Geotherm. Res.* **2016**, *322*, 212–224. [CrossRef]

27. Salvini, R.; Mastrorocco, G.; Esposito, G.; Di Bartolo, S.; Coggan, J.; Vanneschi, C. Use of a remotely piloted aircraft system for hazard assessment in a rocky mining area (Lucca, Italy). *Nat. Hazards Earth Syst. Sci.* **2018**, *18*, 287–302. [CrossRef]

28. Vanneschi, C.; Eyre, M.; Francioni, M.; Coggan, J. The Use of Remote Sensing Techniques for Monitoring and Characterization of Slope Instability. *Procedia Eng.* **2017**, *191*, 150–157. [CrossRef]

29. Abellán, A.; Oppikofer, T.; Jaboyedoff, M.; Rosser, N.J.; Lim, M.; Lato, M.J. Terrestrial laser scanning of rock slope instabilities. *Earth Surf. Process. Landf.* **2014**, *39*, 80–97. [CrossRef]

30. Vanneschi, C.; Salvini, R.; Massa, G.; Riccucci, S.; Borsani, A. Geological 3D modeling for excavation activity in an underground marble quarry in the Apuan Alps (Italy). *Comput. Geosci.* **2014**, *69*, 41–54. [CrossRef]

31. Assali, P.; Grussenmeyer, P.; Villemin, T.; Pollet, N.; Viguier, F. Surveying and modeling of rock discontinuities by terrestrial laser scanning and photogrammetry: Semi-automatic approaches for linear outcrop inspection. *J. Struct. Geol.* **2014**, *66*, 102–114. [CrossRef]

32. Hofierka, J.; Gallay, M.; Bandura, P.; Šašak, J. Identification of karst sinkholes in a forested karst landscape using airborne laser scanning data and water flow analysis. *Geomorphology* **2018**, *308*, 265–277. [CrossRef]

33. Mastrorocco, G.; Salvini, R.; Vanneschi, C. Fracture mapping in challenging environment: A 3D virtual reality approach combining terrestrial LiDAR and high definition images. *Bull. Eng. Geol. Environ.* **2018**, *77*, 691–707. [CrossRef]

34. Vanneschi, C.; Eyre, M.; Venn, A.; Coggan, J.S. Investigation and modeling of direct toppling using a three-dimensional distinct element approach with incorporation of point cloud geometry. *Landslides* **2019**, *16*, 1453–1465. [CrossRef]

35. Locat, J.; Locat, A.; Locat, P.; Robitaille, D.; Turmel, D.; Jaboyedoff, M.; Demers, D.; Oppikofer, T. Use of terrestrial laser scanning for the characterization of retrogressive landslides in sensitive clay and rotational landslides in river banks. *Can. Geotech. J.* **2009**, *46*, 1379–1390.

36. Baltensweiler, A.; Walthert, L.; Ginzler, C.; Sutter, F.; Purves, R.S.; Hanewinkel, M. Terrestrial laser scanning improves digital elevation models and topsoil pH modelling in regions with complex topography and dense vegetation. *Environ. Model. Softw.* **2017**, *95*, 13–21. [CrossRef]

37. Hancock, S.; Anderson, K.; Disney, M.; Gaston, K.J. Measurement of fine-spatial-resolution 3D vegetation structure with airborne waveform lidar: Calibration and validation with voxelised terrestrial lidar. *Remote Sens. Environ.* **2017**, *188*, 37–50. [CrossRef]

38. Kim, D.H.; Balasubramaniam, A.S.; Gratchev, I. Application of photogrammetry and image analysis for rock slope investigation. *Geotech. Eng.* **2018**, *49*, 49–56.

39. Salvini, R.; Mastrorocco, G.; Seddaiu, M.; Rossi, D.; Vanneschi, C. The use of an unmanned aerial vehicle for fracture mapping within a marble quarry (Carrara, Italy): Photogrammetry and discrete fracture network modelling. *Geomat. Nat. Hazards Risk* **2017**, *8*, 34–52. [CrossRef]

40. Tufarolo, E.; Vanneschi, C.; Casella, M.; Salvini, R. Evaluation of camera positions and ground points quality in a gnss-nrtk based uav survey: Preliminary results from a practical test in morphological very complex areas. *ISPRS-Int. Arch. Photogramm. Remote Sens. Spat. Inf. Sci.* **2019**, *42*, 637–641. [CrossRef]

41. Kasprak, A.; Bransky, N.D.; Sankey, J.B.; Caster, J.; Sankey, T.T. The effects of topographic surveying technique and data resolution on the detection and interpretation of geomorphic change. *Geomorphology* **2019**, *333*, 1–15. [CrossRef]

42. Al-Halbouni, D.; Holohan, E.P.; Saberi, L.; Alrshdan, H.; Sawarieh, A.; Closson, D.; Walter, T.R.; Dahm, T. Sinkholes, subsidence and subrosion on the eastern shore of the Dead Sea as revealed by a close-range photogrammetric survey. *Geomorphology* **2017**, *285*, 305–324. [CrossRef]

43. Spetsakis, M.; Aloimonos, J.Y. A multi-frame approach to visual motion perception. *Int. J. Comput. Vis.* **1991**, *6*, 245–255. [CrossRef]

44. Gallup, D.; Frahm, J.-M.; Mordohai, P.; Yang, Q.; Pollefeys, M. Real-Time Plane-Sweeping Stereo with Multiple Sweeping Directions. In Proceedings of the 2007 IEEE Conference on Computer Vision and Pattern Recognition, Minneapolis, MN, USA, 17–22 June 2007; pp. 1–8.

45. Goesele, M.; Snavely, N.; Curless, B.; Hoppe, H.; Seitz, S.M. Multi-view stereo for community photo collections. In Proceedings of the IEEE International Conference on Computer Vision, Minneapolis, MN, USA, 17–22 June 2007.

46. Jancosek, M.; Shekhovtsov, A.; Pajdla, T. Scalable multiview stereo. In Proceedings of the IEEE Workshop on 3D Digital Imaging and Modeling, Kyoto, Japan, 27 September 2009; p. 8.

47. Fanos, A.M.; Pradhan, B. Laser Scanning Systems and Techniques in Rockfall Source Identification and Risk Assessment: A Critical Review. *Earth Syst. Environ.* **2018**, *2*, 163–182. [CrossRef]

48. Kromer, R.; Hutchinson, D.J.; Lato, M.; Gauthier, D.; Edwards, T. Managing rockfall risk through baseline monitoring of precursors using a terrestrial laser scanner. *Can. Geotech. J.* **2017**, *54*, 953–967. [CrossRef]

49. Abellan, A.; Derron, M.-H.; Jaboyedoff, M. "Use of 3D Point Clouds in Geohazards" Special Issue: Current Challenges and Future Trends. *Remote Sens.* **2016**, *8*, 130. [CrossRef]

50. Buill, F.; Núñez-Andrés, M.A.; Lantada, N.; Prades, A. Comparison of Photogrammetric Techniques for Rockfalls Monitoring. *IOP Conf. Ser. Earth Environ. Sci.* **2016**, *44*, 42023. [CrossRef]

51. Wilkinson, M.W.; Jones, R.R.; Woods, C.E.; Gilment, S.R.; McCaffrey, K.J.W.; Kokkalas, S.; Long, J.J. A comparison of terrestrial laser scanning and structure-frommotion photogrammetry as methods for digital outcrop acquisition. *Geosphere* **2016**, *12*, 1865–1880. [CrossRef]

52. Sarro, R.; Riquelme, A.; García-Davalillo, J.C.; Mateos, R.M.; Tomás, R.; Pastor, J.L.; Cano, M.; Herrera, G. Rockfall Simulation Based on UAV Photogrammetry Data Obtained during an Emergency Declaration: Application at a Cultural Heritage Site. *Remote Sens.* **2018**, *10*, 1923. [CrossRef]

53. Saroglou, C.; Asteriou, P.; Zekkos, D.; Tsiambaos, G.; Clark, M.; Manousakis, J. UAV-based mapping, back analysis and trajectory modeling of a coseismic rockfall in Lefkada island, Greece. *Nat. Hazards Earth Syst. Sci.* **2018**, *18*, 321–333. [CrossRef]

54. Salvini, R.; Francioni, M. Geomatics for slope stability and rock fall runout analysis: A case study along the alta tambura road in the apuan alps (Tuscany, Italy). *Ital. J. Eng. Geol. Environ.* **2013**, 481–492.

55. Salvini, R.; Francioni, M.; Riccucci, S.; Bonciani, F.; Callegari, I. Photogrammetry and laser scanning for analyzing slope stability and rock fall runout along the Domodossola-Iselle railway, the Italian Alps. *Geomorphology* **2013**, *185*, 110–122. [CrossRef]

56. Žabota, B.; Repe, B.; Kobal, M. Influence of digital elevation model resolution on rockfall modelling. *Geomorphology* **2019**, *328*, 183–195. [CrossRef]

57. Frattini, P.; Crosta, G.; Carrara, A.; Agliardi, F. Assessment of rockfall susceptibility by integrating statistical and physically-based approaches. *Geomorphology* **2008**, *94*, 419–437. [CrossRef]

58. Dussauge-Peisser, C.; Helmstetter, A.; Grasso, J.-R.; Hantz, D.; Desvarreux, P.; Jeannin, M.; Giraud, A. Probabilistic approach to rock fall hazard assessment: Potential of historical data analysis. *Nat. Hazards Earth Syst. Sci.* **2002**, *2*, 15–26. [CrossRef]

59. Colantoni, P.; Menichetti, M.; Savelli, D.; Tramontana, M. *Carta Geologico e Geomorfologica*; University of Urbino: Urbino, Italy, 2012.

60. Pichezzi, R.M. *Carta Geologica d'Italia a Scala 1:50.000-Foglio 280 (Fossombrone)*; ISPRA Serv. Geologico d'Italia: Roma, Italy, 2016.

61. Pergolini, C.; Farina, D. Gli aspetti geografici, geologici, paleontologici. In *I Monti del Furlo*; Regione Marche, Comunità Montane Metauro, Alto e Medio Metauro, Catria e Nerone: Ancona, Italy, 1990; pp. 11–36.

62. Centamore, E.; Deiana, G. *Carta Geologica delle Marche a Scala 1:250.000*; Università di Camerino e Regione Marche, LAC: Firenze, Italy, 1986.

63. Del Prete, M. *Progetto per la Realizzazione di reti di Protezione ed Altre Opera e Attività Complementari Sulle Pareti Rocciose Soggette a Crollo Sovrastanti la via Flaminia, nel Tratto Compreso tra la Galleria Romana e le Località Furlo*, Technical Report; Amministrazione Provinciale di Pesaro e Urbino: Pesaro, Italy, 2012; p. 11.

64. Centamore, E.; Cantalamessa, G.; Micarelli, A.; Potetti, M.; Berti, D.; Bigi, S.; Morelli, C.; Ridolfi, M. Stratigrafia e analisi di facies dei depositi del Miocene e del pliocene inferiore dell'avanfossa marchigiano-abruzzese e delle zone limitrofe. *Stud. Geol. Camerti* **1991**, *11*, 121–131.

65. Besl, P.; McKay, H. A method for registration of 3-D shapes. *IEEE Trans. Pattern Anal. Mach. Intell.* **1992**, *14*, 239–256. [CrossRef]

66. Cima, V.; Carroccio, M.; Maseroli, R. Corretto utilizzo dei Sistemi Geodetici di Riferimento all'interno dei software GIS. In Proceedings of the Atti 18a Conferenza Nazionale ASITA, Firenze, Italy, 14–16 October 2014; pp. 359–363.
67. Hansman, R.J.; Ring, U. Workflow: From photo-based 3-D reconstruction of remotely piloted aircraft images to a 3-D geological model. *Geosphere* **2019**, *15*, 16. [CrossRef]
68. Chudley, T.R.; Christoffersen, P.; Doyle, S.H.; Abellan, A.; Snooke, N. High-accuracy UAV photogrammetry of ice sheet dynamics with no ground control. *Cryosphere* **2019**, *13*, 955–968. [CrossRef]
69. Bieniawski, Z.T. Engineering Classification of Jointed Rock Masses. *Civ. Eng. South. Afr.* **1973**, *15*.
70. Bieniawski, Z.T. *Engineering Rock Mass Classifications: A Complete Manual for Engineers and Geologists in Mining, Civil, And Petroleum Engineering*; Wiley-Interscience Publication: New York, NY, USA, 1989; p. 272.
71. Hoek, E.; Brown, E. Practical estimates of rock mass strength. *Int. J. Rock Mech. Min. Sci.* **1997**, *34*, 1165–1186. [CrossRef]
72. Thiele, S.T.; Grose, L.; Samsu, A.; Micklethwaite, S.; Vollgger, S.A.; Cruden, A.R. Rapid, semi-automatic fracture and contact mapping for point clouds, images and geophysical data. *Solid Earth* **2017**, *8*, 1241–1253. [CrossRef]
73. Markland, J.T. *A Useful Technique for Estimating the Stability of Rock Slopes When the Rigid Wedge Sliding Type of Failure Is Expected*; Tylers Green, H.W., Ed.; Imperial College of Science and Technology: London, UK, 1972; Volume 19.
74. Gigli, G.; Casagli, N. Semi-automatic extraction of rock mass structural data from high resolution LIDAR point clouds. *Int. J. Rock Mech. Min. Sci.* **2011**, *48*, 187–198. [CrossRef]
75. Ferrero, A.M.; Forlani, G.; Roncella, R.; Voyat, H.I. Advanced geostructural survey methods applied to rock mass characterization. *Rock Mech. Rock Eng.* **2009**, *42*, 631–665. [CrossRef]
76. Mavrouli, O.; Corominas, J.; Jaboyedoff, M. Size Distribution for Potentially Unstable Rock Masses and In Situ Rock Blocks Using LIDAR-Generated Digital Elevation Models. *Rock Mech. Rock Eng.* **2015**, *48*, 1589–1604. [CrossRef]
77. Chen, N.; Kemeny, J.; Jiang, Q.; Pan, Z. Automatic extraction of blocks from 3D point clouds of fractured rock. *Comput. Geosci.* **2017**, *109*, 149–161. [CrossRef]
78. Kazhdan, M.; Bolitho, M.; Hoppe, H. Poisson surface reconstruction. In Proceedings of the Fourth Eurographics Symposium on Geometry Processing, Sardinia, Italy, 26–28 June 2006; pp. 61–70.
79. Ministero dei Lavori Pubblici. *NTC-Norme Tecniche per le Costruzioni*; Repubblica Italiana: Roma, Italy, 2018.
80. Gallego, J.G.; Fonseca, R.J.L. Optimization criteria for using dynamic rockfall protection systems. In Proceedings of the Second Half Century of Rock Mechanics, Lisbon, Portugal, 9–13 July 2007; Riberio, L., Sousa Olalla, C., Grossmann, N., Eds.; pp. 637–640.
81. Dorren, L.K.A. *Rockyfor3d (V5.1) Revealed–Transparent Description of the Complete 3d Rockfall Model*; ecorisQ Association: Geneva, Switzerland, 2012.
82. Dorren, L.K.A.; Heuvelink, G.B.M. Effect of support size on the accuracy of a distributed rockfall model. *Int. J. Geogr. Inf. Sci.* **2004**, *18*, 595–609. [CrossRef]
83. Netti, T.; Castelli, M.; De Biagi, V. Effect of the Number of Simulations on the Accuracy of a Rockfall Analysis. *Procedia Eng.* **2016**, *158*, 464–469. [CrossRef]

International Journal of
Geo-Information

Article

The Suitability of UAS for Mass Movement Monitoring Caused by Torrential Rainfall—A Study on the Talus Cones in the Alpine Terrain in High Tatras, Slovakia

Rudolf Urban [1,*], Martin Štroner [1], Peter Blistan [2], Ľudovít Kovanič [2], Matej Patera [2], Stanislav Jacko [3], Igor Ďuriška [3], Miroslav Kelemen [4] and Stanislav Szabo [4]

[1]   Department of Special Geodesy, Faculty of Civil Engineering, Czech Technical University in Prague, Thákurova 7, 166 29 Prague, Czech Republic

[2]   Institute of Geodesy, Cartography and Geographical Information Systems, Faculty of Mining, Ecology, Process Control and Geotechnology, Technical University Kosice, Park Komenského 19, 04001 Košice, Slovakia

[3]   Institute of Geosciences, Faculty of Mining, Ecology, Process Control and Geotechnology, Technical University Kosice, Park Komenského 15, 04001 Košice, Slovakia

[4]   Department of Flight Training, Faculty of Aeronautics, Technical University of Kosice, Ramp 7, Kosice 04121, Slovakia

*   Correspondence: rudolf.urban@fsv.cvut.cz

Received: 8 May 2019; Accepted: 20 July 2019; Published: 24 July 2019

Abstract: The prediction of landslides and other events associated with slope movement is a very serious issue in many national parks around the world. This article deals with the territory of the Malá Studená Dolina (Little Cold Valley, High Tatras National Park—Slovakia), where there are extensive talus cones, through which seasonally heavy hiking trails lead. In the last few years particularly, there have been frequent falls and landslides in the mountainous environment, which also caused several fatal injuries in 2018. For the above reasons, efforts are being made to develop a methodology for monitoring the changes of the talus cones in this specific alpine area, to determine the size, speed, and character of the morphological changes of the soil. Non-contact methods of mass data collection (laser scanning with Leica P40 and aerial photogrammetry with unmanned aerial system (UAS) DJI Phantom 4 Pro) have been used. The results of these measurements were compared and the overall suitability of both methods for measurement in such terrain evaluated. The standard deviation of the difference of surface determination (represented by the point cloud) is about 0.03 m. As such accuracy is sufficient for the purpose of monitoring talus cones and the use of UAS is easier and associated with lower risk of damage of expensive equipment, we conclude that this method is more suitable for mapping and for repeated monitoring of such terrain. The properties of the outputs of the individual measurement methods, the degree of measurement difficulty and specific measurement conditions in the mountainous terrain, as well as the economy of the individual methods, are discussed in detail

Keywords: monitoring; georelief; geohazards; talus cones; UAS; TLS; SfM; torrential rainfall

---

1. Introduction

The alpine terrain with its exposed georelief and climatic conditions supports a wide range of natural processes with various morphodynamic phenomenons. Various influences, including water, sunshine, and temperature changes, on the georelief of the high mountains are reflected in the spatial incidence and intensity of morphological processes such as water-, snow-, and frost-induced processes, solifluction, deflation, etc. Many of these processes occur in the highest positions of the ridges and peaks; others affect the slopes of the valleys and also reach their foot. Nowadays, in view of climate change, monitoring of the dynamics of georelief evolution and its changes is a frequently discussed topic. The choice of an appropriate data collection method from the point of view of the accuracy, technical complexity, cost-efficiency, and overall suitability is of utmost importance. The use of modern methods of non-contact survey, including digital photogrammetry performed by unmanned aerial system (UAS) carriers or terrestrial laser scanning (TLS), appear to bring significant advantages over the traditional methods of geodesy and cartography, such as measurements using total stations [1]. These traditional procedures are not suitable for detailed monitoring of flat or spatially irregular formations as signaling and subsequent measurements of individual points is unjustifiable in terms of time and money (e.g., [2,3]).

Changes in the position of objects, the shape of the terrain, the morphology of landslides or in anything else in the landscape can be determined by the stage method of measurement—i.e., by taking measurements of the area of interest in various time points. Following the baseline measurement, another one is taken either after a pre-set time period or after a significant event (for example, torrential rain) and the differences between both measurements are analyzed.

As photogrammetric methods (or 3D laser scanning) in principle do not directly relate to a global coordinate system (georeferencing), the use of a certain number of ground control points (GCPs) is required by terrestrial geodesy (with georeferencing) or, more often, by GNSS (global navigation satellite systems), with subsequent transformation into terrestrial position and elevation systems.

UASs have been previously used for many applications, including monitoring of changes in the morphology of a volcano [4], landslides [5,6], dam and riverbed erosion [7–9], slow landslides [10], risks associated with surface mining [11], slope stability in the vicinity of railways [12], or speed of glacier movement [13,14].

Commonly achievable accuracy reported in most studies is 5–10 cm. The terrain monitored in those studies is usually smooth, practically free of vegetation or with sparse vegetation only. A less frequently studied problem is the monitoring of areas covered by vegetation, which makes it difficult to determine changes. We rarely read about monitoring changes in terrain morphology and geological phenomena such as rock blocks, glacier moraine, talus cones, and scree cones, located in hard-to-reach but visited mountain areas.

In this paper we present a procedure for documenting and monitoring the talus cone stability, which is a typical geological phenomenon in alpine areas, using the SfM (Structure from Motion) method based on UAS imaging. The National Park of High Tatras is probably the most visited park in Slovakia and the instability of talus cones during times of of torrential rains thus may present a significant danger to the tourists in their vicinity. To be able to monitor the hazards presented by those cones, however, it is necessary to develop inexpensive but effective methods for detailed description of the surface that could be systematically repeated over time. As the use of traditional geodesy methods is unsuitable for this task, the presented paper aims at developing a method for effective and systematic monitoring using UAS. A significant problem of documentation in the alpine environment is posed by the character of the terrain, which is inclined with a slope usually between 20° and 60°, often very rugged, consisting essentially of larger or smaller boulders ranging in sizes from 10 cm to 1 m (often with the occurrence of isolated stone blocks of several meters). For these reasons, more extensive terrain reconnaissance and consistent planning of the flight is necessary so that a continuous surface can be reconstructed from the data. Due to the inclination and ruggedness of the terrain, it is not practically possible to capture the surface by traditional methods of geodesy, which calls for the use of methods of mass data collection. For this reason, terrestrial 3D laser scanning was chosen as the most appropriate comparative method to verify the quality of the digital terrain model obtained from UAS photogrammetry. The goal of the presented research was to analyze and evaluate the overall suitability (accuracy, quality, laboriousness, efficiency, and usability of the tested methodology) of low-cost UAS photogrammetry for the needs of monitoring selected geohazards (such as rock collapses, glacial moraine, stone fields, and talus cones) located in very specific, rugged, and inclined terrains.

The selected area is typical for the medium-high Alpine-type mountains, such as the above-mentioned Alps or Carpathians, on the example of a talus cone located in the Little Cold Valley in High Tatras.

2. Study Area

The area of interest for monitoring changes of georelief by using UAS photogrammetry was a selected area of High Tatras (in the northern part of Slovakia on the border with Poland), namely the part of the Little Cold Valley (Figure 1).

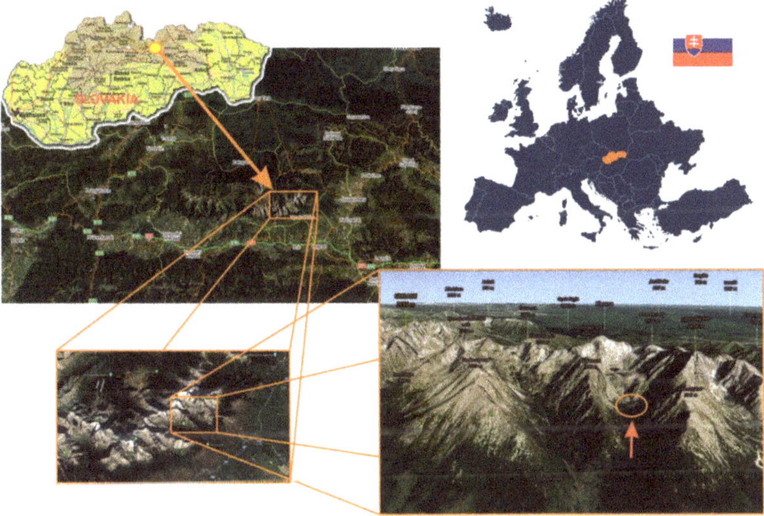

Figure 1. Location—Little Cold Valley in the High Tatras, Slovakia.

The whole territory of the Tatras belongs to the Tatra National Park—TANAP. The area of its own territory is 73,300 ha, and the protection zone is 39,800 ha. TANAP is the highest mountain group in the Carpathian arch. The most valuable part of the area is the 55 km long and 17 km wide geomorphological unit of the Tatras.

The evolution and morphostructural shape of the High Tatras is linked with polyphase tectonic, initialized by Alpine-Carpathian collisional processes that started during the transition from Mesozoic to Cenozoic period and culminated in the Oligocene-Early Miocene by vertical uplift of the mountain range. This robust uplift uncovers Variscan crystalline basement of the Tatric Unit. The younger alpine supra structures such as the Fatric and Hronic Units are conserved only in the marginal parts of the mountains. The Tatric Unit is built by pre-Mesozoic rocks where S-type granodiorites dominate [15,16], respectively medium-grade mica schists. The recent topography of the High Tatras relates to Pleistocene glaciation [17–19] during which 55 glacier systems were created [20] based on the distribution glacial erosional and depositional landforms. The high alpine surface is modified by the processes of weathering, erosion, gravity, and debris flow transport and accumulation [21–25] of granodiorite debris on the sides of the valleys. The debris development is accelerated by the presence of weathering prone mica and an irregular network of faults and joints. The deposits on the talus slopes are positioned on the top of moraine sediments or of floodplain deposits of recent mountain streams that overlap moraine sediments [26] and are built by either proluvial (alluvial), deluvial, or proluvial-deluvial cones. The cones have a typical construction consisting of source, transport, and deposition area. Water transports material in migrating distribution channels, which are often characterized by their "paternoster" structure, where large debris boulders block the channel transport of smaller materials. During massive rainfall events, the strong water inflow causes the boulders to be released. In the case that accumulation areas of talus slopes are on the floodplain of a recent mountain stream, the deposits are further eroded and carried away from the valley by the stream during the increased water period. The inactive areas of talus slopes are afforested with dwarf mountain pine.

The high alpine surface of the High Tatras is typical for younger morphostructures of the Alpine and Carpathian belt. It is characterized by ragged, steep, high relief valleys filled by regolith. These valleys are developed in glacial or glacifluvial conditions. Recently, several studies have shown that the modern evolution of European high alpine surfaces is closely related to climate change [27–29], especially to temperature anomalies and the increase of intense rainfall events.

The character of the climate is a transitional character between oceanic western European and continental climate, where the cold climatic region of the alpine type prevails. According to the Köppen Climate Classification System, the area belongs to the ET (tundra) climate region [30]. The snow cover stays for 200–250 days a year—in firn areas the whole year. Even in summer, precipitations are often in the form of snow in altitudes above 2000 m. More than 200 days a year, the average daily temperature is below 0 °C.

In the whole area of interest, there are several geomorphological formations of the character of rubble streams, which are increasingly active due to enormous rainfall in the last 10 years. They can potentially be included in geohazards because they are located in the tourist-frequented valleys of the High Tatras (frequented high mountain hiking trails often pass through these places).

The biggest mass movements and talus cones are in Tatras located at the end of trough valleys because these have the highest walls modeled by glaciers. Here, often, talus cones are joined together with alluvial cones to form combined-type alluvial talus cones. The features of such a combined cone are also present in the highest talus cone in the Tatras located at the end of the Kolová Valley under the NE wall of the Kolový Peak. It is 430 m high with water-flooded foot reaching the Kolové tarn (1565 m above sea level). Major talus cones can be also found in [26]:

- The Small Cold Valley, 400 m high;
- Mengusovska Valley below Satan Peak with a height of 400 m;
- East of Vareškove tarn with a height of 330 m.

The author considers enormously high talus cones to be those that exceed 200 m. The same types of talus cones are observable in the Alps, and respectively in all alpine-type mountains.

Mass Movement as a Significant Phenomenon of the Alpine Type Mountains

According to [31], under the term mass movement, we understand the geomorphological process as well as the forms formed under the influence of short-term atmospheric precipitation in the talus material. Precipitation of more than 1 mm·min^{-1} and yields of 40 mm/day accelerate the movement of debris material on the slopes in the Tatras. While in the Alps mass movements of 400,000–500,000 m^3 have been recorded, in the Tatras it is up to 25,000 m^3, and in Scotland and Scandinavia only 100–350 m^3 [31,32]. Their size is determined by both morphometric and lithological properties. In this case, it is useful to know the debris cover. Statistical measurements show that hourly rainfall intensities of 60–80 mm occur in the Tatras with a probability of 1%—i.e., once every hundred years. However, an intensity of 40 mm·h^{-1} may occur with a probability of 10%—i.e., once every 10 years. Over the past seven years, the intensity of extreme rainfall has risen compared to the earlier measurements carried out by [33]. As reported by [31], since 1995 in the Tatra Mountains the transition to a more humid climate with the development of fluvial processes is observed, i.e., manifestations of overland flow and concentrated surface runoff in the form of rain rills and washout channels. At present, we observe very active talus cones (seasonal changes are visible by the naked eye) in the area of the Mlynická Valley, Mengusovská Valley, Maple Valley, Great and Little Cold Valley.

Negative manifestations of massive mass movement, such as the destruction of the tourist track located below the facewall of the Lomnický Peak, are well apparent in the Little Cold Valley. The source area starts at an altitude of approximately 2200 m and the width of mass movement was approximately

1250 m in 2014, which is 50 m more than in 2004. Its width in the accumulation zone also increased from the original 5–12 m to 96 m. The same changes in the morphology of mass movement, which is the subject of our research, lie in the close proximity to a previously mentioned mass movement with the source area below Lomnicky Peak, documented in the images from 2010 and 2016 (Figure 2). There is also an increase in the length and width of the mass movement over the years.

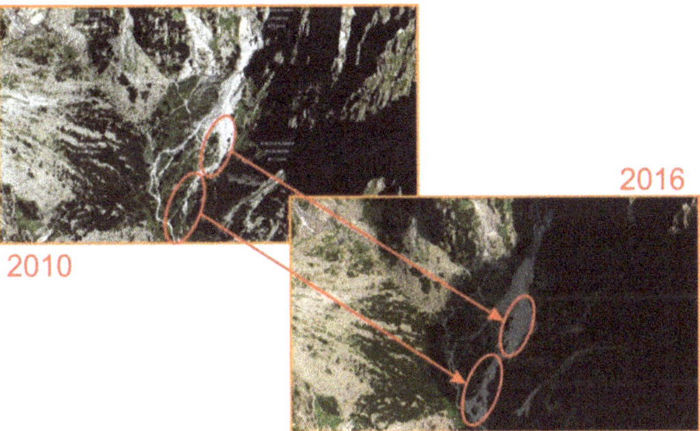

Figure 2. Visible changes in the monitored talus cone in the Little Cold Valley, High Tatras, Slovakia. Aerial photos from 2010 and 2016.

3. Methods and Instruments

Long-term geodetic monitoring of selected geohazards, such as rock collapses, glacier moraine, stone fields, mass movement, using conventional geodetic methods and instruments (total station or 3D laser scanner) is a technically and physically very demanding process. To verify the methodology and the possibility of monitoring geohazards and especially stone fields, rock collapses, and rubble cones, located in very specific, rugged, and inclined terrains in the medium-high Alpine-type mountains, the Small Cold Valley in the High Tatras, Slovakia was chosen. This valley was selected as a typical alpine landscape, with typical glacial morphology and occurrence of typical geologic phenomena, including the occurrence of Alpine-similar vegetation. At the end of the valley, there is a rubble stream (below the walls of Lomnicky Peak, Figure 3).

This stream was monitored in this research by methods of mass data collection—tested method low-cost UAS photogrammetry. Trusted reference measurements were performed by 3D terrestrial laser scanning. We used the approach for the UAS data collection and process workflow for the needs of georelief modelling described in [34] (Figure 4), carried out in five basic steps:

1. Terrain reconnaissance;
2. Preparatory work and pre-flight preparation;
3. Photogrammetric data collection;
4. Processing of aerial images;
5. Creation of terrain model.

During processing, Agisoft Photoscan uses camera callibration parameters (elements of interior orientation) based besides ground control points also on key points. Hence, the calibration is based on a high number of points for each individual image (thousands: in the settings, 40,000 points per image were set as a maximum). In Figure 4, Step 2, pre-flight calibration of the camera is shown as an option. In our study, we used a post-processing camera calibration feature in Agisoft Photoscan.

Figure 3. A documented active talus cone in the Little Cold Valley, High Tatras, Slovakia.

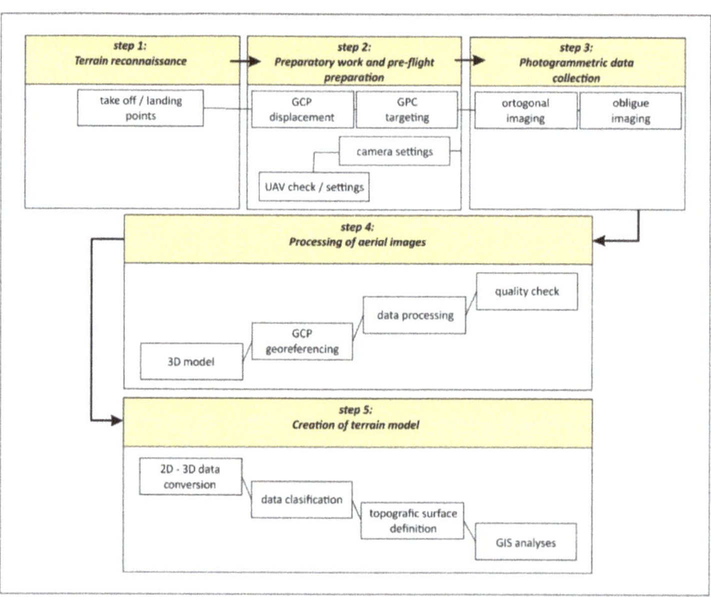

Figure 4. Example of data collection methodology using unmanned aerial system (UAS) for modeling georelief, according to [33].

It was necessary to build a precise geodetic network for research purposes. The geodetic network was stabilized by four copper studs (affixed into the rock using Fischer chemical mortar) and three reflective targets (reflective foils) glued with a special adhesive (Pattex Chemoprén Universal) to suitable flat rock surfaces or large stone blocks (see Figure 5 for locations). There were also three temporary points measured using GNSS RTK (Real Time Kinematic) method with connection to the Leica SmartNet network, which formally served to connect a geodetic network to local Datum of Uniform Trigonometric Cadastral Network and height above sea level (Baltic Vertical Datum—After Adjustment). GNSS measurements were made by a Leica GPS900cs receiver. Temporary GNSS points were used only to determine the spatial position of the total station, and then everything was measured by the spatial polar method (geodetic network, UAS ground control points, scanner ground control points). Geodetic measurements were performed by a total station of Leica TS02 (Figure 6, left; Table 1) with an angular accuracy of 7″ (0.0020 gon) and a length accuracy of 2 + 2 ppm. The measurement was performed using two telescope faces from a single station. The spatial accuracy (standard deviation) of the individual points was below 10 mm. During measurements, the spatial position of the total station was checked regularly.

Figure 5. Geodetic network (crosses mark the points of geodetic network and the arrow indicates the position of the total station. Source of map: maps.google.com).

Figure 6. Leica TS02 total station (left), DJI Phantom 4 Pro (middle), and 3D scanner Leica P40 (right).

The UAS DJI Phantom 4 Pro (Figure 6, middle) equipped with a 5472 × 3648 pixel camera was used to capture the images of the talus cone. Altogether, 1389 images were taken in several flights from an average height of 35 m above ground, with 1 pixel therefore representing approximately 0.01 m. The automatic camera mode with fixed ISO (100) was used during the flight (ISO 100, Shutter 1/60 to 1/800, F/3.5 − F/7.1). The total flight time was about 3 h.

Table 1. Leica TS02 characteristics.

Angle Measurements (Hz, V)	
Accuracy	7"
Compensation	Angular compensation
Length measurement with a prism	
Reach	3.500 m
Accuracy	Accurate+: 1.5 mm + 2.0 ppm, Accurate Fast: 3.0 mm + 2.0 ppm, Tracking: 3.0 mm + 2.0 ppm
Distance measurement without a prism	
Range	>400 m
Accuracy	2 mm + 2 ppm
Operation	
Operating temperature	−20 to 50 °C (−4 to 122 °F) Arctic version −35 to 50 °C (−31 to 122 °F)

The ground control points were made of fibreboard 0.3 × 0.3 m with a black and white target. The GCPs were prevented from dislocation by being weighted down using stones collected on site, the positions are shown in Figure 7. The image acquisition was performed in four flights due to battery endurance. The first flight was manually piloted, with the aim of maintaining a stable height above the terrain. Due to the difficulties with spatial orientation on the study site, it was guided with the help of a co-worker in a way that allowed proper coverage of the study area. The three remaining flights used an autopilot and pre-programmed flight paths (pre-programming was performed in the Pix4D software). The aim was to achieve the best possible geometry of camera positions for subsequent 3D modelling. The flight height was set to 30 m above the take-off point as the Pix4D software does not support different heights in different waypoints. To obtain images covering the entire study area from a similar distance above terrain, the study area was divided into three parts, each covering an area with a similar altitude. All flight missions used a double grid pattern; however, the last flight had to be interrupted due to a sudden worsening of meteorological conditions (strong gusts of wind): therefore, the bottom part of the area was not covered using a double grid pattern (see Figure 10, left).

Figure 7. Location of ground control points (red dots) and installation of a ground control point.

Terrestrial laser scanning was performed with the Leica P40 (Figure 6, right; Table 2), which has a two-axis compensator, 1.2 mm + 10 ppm accuracy of measured distance, 8″ angular accuracy (0.0025 gon), 3 mm/50 m and 6 mm/100 m for 3D point accuracy, a scan speed of up to 1 million points per second and 360 × 270° field of view. For each scanner standpoint, at least three temporary control points (black and white targets that can be scanned automatically after selection) were measured by the total station to register the resulting point clouds. The scanner resolution was set to 12 mm/10 m with a range of 120 m.

Table 2. Leica P40 3D scanner characteristics.

Main Specification	
3D scanner characteristics	compact, pulse, dual-axis compensator
Accuracy	
Range	1.2 mm + 10 ppm
Angular	8" horizontal; 8" vertical
3D position	3 mm at 50 m; 6 mm at 100
Target acquisition	2 mm standard deviation at 50 m
Distance Measurement System	
Type	Ultra-high speed time-of-flight enhanced by Waveform Digitising (WFD) technology
Range and reflectivity	Minimum 0.4 m, 270 m@34%; 120 m@8%
Scan rate	Up to 1,000,000 points per sec
Field of view	H – 360° (max.); V – 290° (max.)
Range noise	0.4 mm rms at 10 m, 0.5 mm rms at 50 m

A total of 25 standpoints (see Figure 8) were used, with a scanning time of about 12 h, which amounted (including transport and movement in the difficult terrain) to two working days—it was difficult indeed to carry a scanner weighing around 28 kg (including batteries and a protective box) around and through the stone fields.

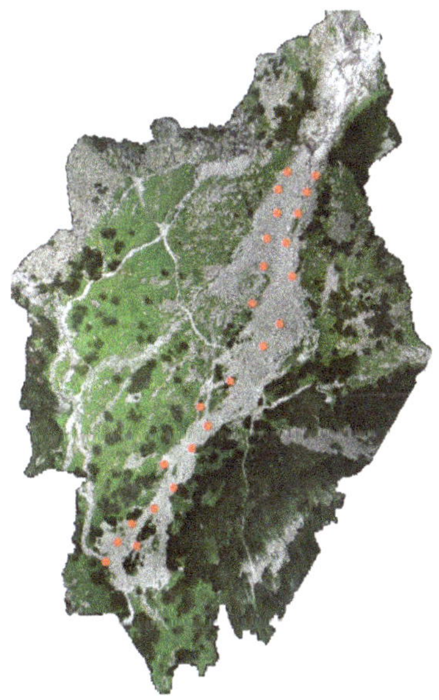

Figure 8. Leica P40 standpoints.

4. Processing of Measured Data

3D data from the laser scanner was only processed by transforming individual positions into a common coordinate system by ground control points, which were determined by the total station measurement with connection to stabilized points of the geodetic network. There was no need for further processing. The resulting point cloud contained 597 million points. The scan registration accuracy was below 3 mm in all stations. The mean RMSE (root-mean-square error) was approx. 1.2 mm.

The processing was carried out in the Leica Cyclone program. The resulting data is shown in Figure 9; Figure 9A shows measurements from one standpoint of the instrument, where, due to the character of the terrain (boulders, etc.), there are fundamental obstacles and the data itself is very incomplete. Figure 9B shows the situation after merging all the data for a given area, where the situation is already considerably better, but the terrain coverage is definitely not compact.

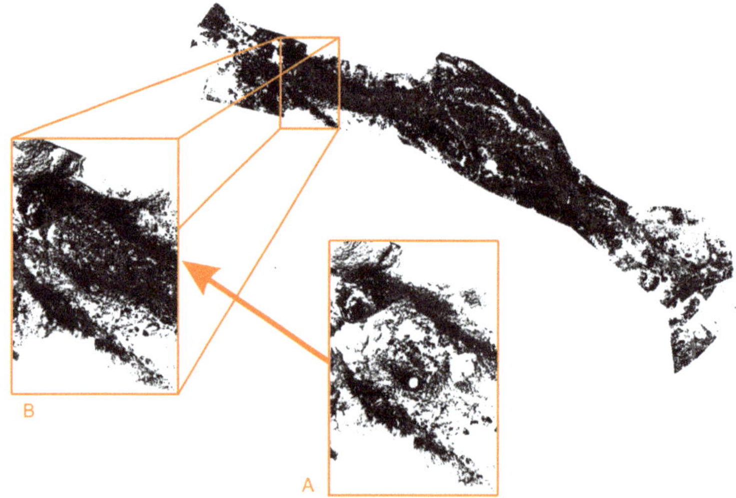

Figure 9. 3D scanning data after registration—whole area of interest. (**A**) Detail of data acquired from one standpoint, and (**B**) detail of complete data.

UAS images were processed in Agisoft PhotoScan ver. 1.2.5. A total of 1389 images were used (Figure 10, left). The processing was very problematic due to the detail level required. The calculation quality was set to "High quality" (where the original image resolution is used) when aligning images even when generating point clouds. A total of seven computers and a server in a common network solution were used for the calculation. The alignment was performed in a bulk for all frames, after which, due to computational demands, the area had to be divided into nine parts for separate generation of point clouds and the data was subsequently merged. The areas were chosen gradually so that they always remain unchanged at the two territorial boundaries. A total of 261 million points were obtained after the data was merged (Table 3).

Table 3. Numbers of points in the acquired point clouds.

Point Cloud by Method	Number of Points
SfM	261,097,729
TLS—original	597,031,328
TLS—filtered	532,956,824

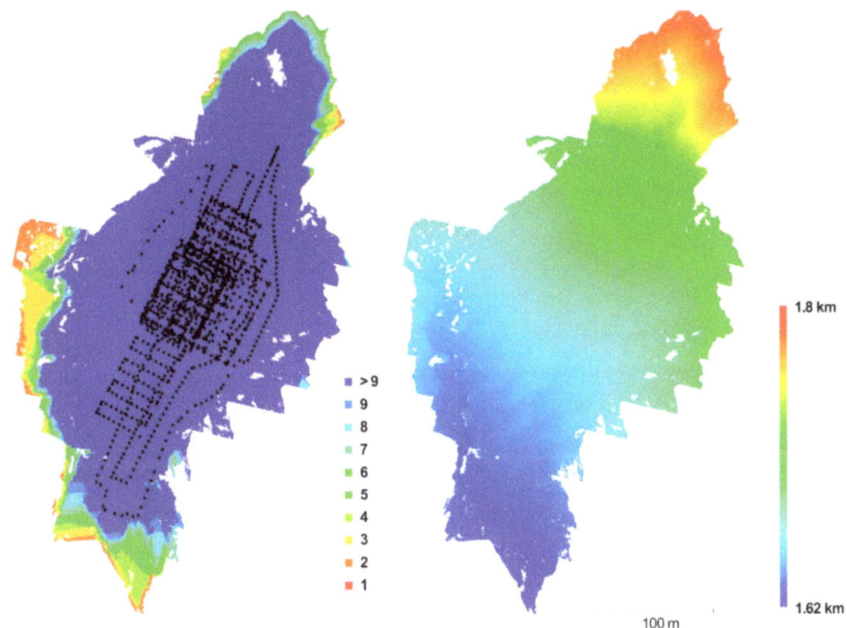

Figure 10. The number of UAS images capturing the area and the flight plan; dots represent camera positions during image acquisition (left) and reconstructed digital elevation model with the color scale shows altitudes (right).

The quality of the photogrammetric model was inspected after the calculation (Table 4). The RMSE on ground control points was less than 0.025 m and the mean overall RMSE was 0.011 m; the RMSE of image coordinates was always less then 0.32 pix and the mean RMSE of coordinates was 0.16 pix. The RMSE values characterize the internal model agreement, not the absolute accuracy of the generated point cloud. The evaluation of the SfM-derived point cloud is performed further on in the paper by a comparison with TLS results. The altitudes of the terrain are shown in Figure 10 (right). The resulting point cloud obtained from the SfM method is shown in Figure 11 depicting an overview and detail of the terrain capture. The detail of terrain capture and virtually non-creeping coverage of the area suggests greater suitability of the SfM method for such specific terrain.

Data comparison was performed in CloudCompare [35]. Due to the discontinuity of 3D laser scanning data, it was necessary to calculate the data using SfM as the reference, and to subsequently project the 3D scan data on it. If done the other way around, photogrammetrically obtained data would show high deviations in the places where 3D scan data is missing simply due to the absence of data. The points representing vegetation (shrubs, grass, and other lower small vegetation) were manually removed from the clouds before the comparison as much as possible, however some small areas of vegetation could not be manually removed.

The point clouds were directly compared by determining the minimum distance of each individual point from the TLS cloud to an irregular triangular network formed between the nearest nine points of the SfM cloud (function Cloud to Cloud, tab Local modelling, Local model option 2D1/2 Triangulation). Only comparison on the vertical axis (the only relevant one for time series analyses) was performed.

The absolute distances, as well as the individual components of this distance in the direction of the X, Y, and in particular Z coordinate axes, were calculated. The height component is very important for the resulting assessment of data quality or their mutual consent. The function "Compute cloud-to-cloud distance" was used, the components of length in the X, Y, Z directions were calculated, and the local surface modeling using a triangular mesh made of the nearest nine points was used. The average distance (systematic shift) and the standard deviation of the differences were also calculated. To get rid of outliers that were to a large degree (besides being inherently present in any measurement) caused by islets of vegetation that could not have been manually removed and by deep shadow areas between stones, we subsequently filtered them out by removing all data where the difference betweeen clouds was higher than 2.5* σ_Z (i.e., removing approx. 1% of the most outlying values), thus creating a filtered TLS cloud (Table 3), and repeated the accuracy evaluation of the filtered cloud as well.

Table 4. The residuals calculated for the ground control points after the bundle block adjustment.

Control Point Number	RMSE (m)	X_{error} (m)	Y_{error} (m)	Z_{error} (m)	Image (pix)
1001	0.022	−0.006	−0.021	−0.002	0.122
1002	0.013	−0.008	0.009	−0.006	0.093
1003	0.011	0.010	−0.005	0.000	0.201
1004	0.010	0.001	0.008	0.006	0.205
1005	0.021	−0.008	0.018	−0.005	0.150
1006	0.011	0.009	−0.005	0.002	0.145
1007	0.005	0.004	−0.002	0.002	0.169
1008	0.006	−0.002	0.005	0.003	0.138
1009	0.002	0.002	−0.002	0.001	0.138
1010	0.006	−0.002	−0.005	0.003	0.128
1011	0.003	0.001	−0.003	−0.002	0.168
1012	0.002	−0.001	−0.002	−0.001	0.150
1013	0.006	−0.006	−0.001	0.002	0.168
1014	0.012	0.005	−0.010	−0.006	0.183
1015	0.003	0.002	−0.002	−0.002	0.227
1016	0.016	0.000	0.016	0.004	0.318
RMSE	0.011	0.005	0.009	0.004	0.159

Figure 11. Data from photogrammetric image processing by SfM method—entire area and detail.

5. Results

By field research on a pre-selected site in the Little Cold Valley in the High Tatras, data capturing the same object of interest—the talus cone—were collected using two methods at the same time. Measurements were taken simultaneously—photogrammetically from images taken via UAS from an approximate flight height of 35 m above terrain, and by terrestrial 3D laser scanning by a device placed on a tripod at approximately 1.5 m above ground. The control points were determined in both cases from the same geodetic network.

5.1. Evaluation of Precision of the Created Point Clouds

The precision of each point of the TLS point cloud related to the standpoint acquired by the 3D scanner is in our case given by the standard deviation in position less than 3 mm (according to the manufacturer's specifications, the measurement distance was shorter than 50 m). The precision of the registration was in our case characterized by the RMSE 1.2 mm. The ground control points precision was 10 mm. Total standard deviation in position of the individual point of the TLS point cloud can then be derived (by the application of the standard deviation propagation law) as 10 mm. Uncertainty added to the results by the TLS measurement itself is therefore negligible in view of the purpose of the monitoring. In contrast, the uncertainty of the SfM method point cloud generation is much higher—the point cloud is, in comparison to the TLS cloud, very noisy, as is generally known. Thus, the 3D scanning method can be considered more accurate and is used as the reference one.

The visualization of the result of the comparison of the point clouds (TLS and UAS) is shown in Figure 12, with various colors representing the differences.

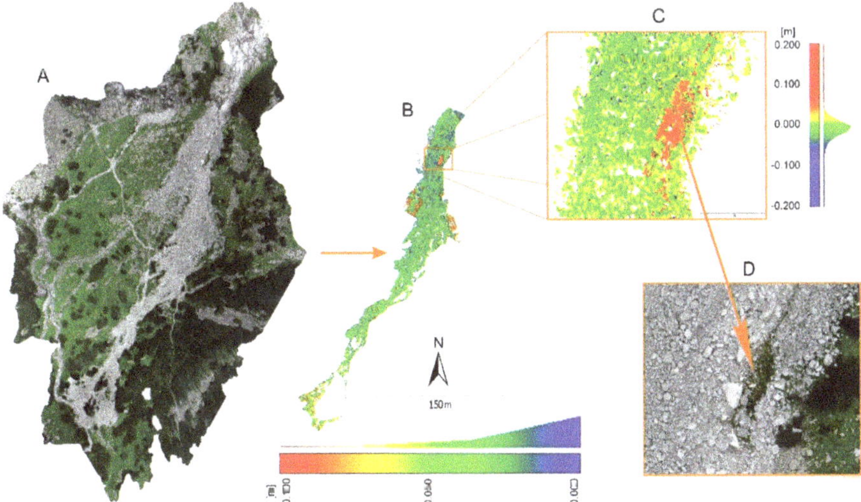

Figure 12. The result of comparing data from terrestrial 3D laser scanning and UAS photogrammetry. (**A**) The whole area of interest (only the talus cone of which was analyzed, the rest containing vegetation was manually removed); (**B**) comparison of the unfiltered clouds showing absolute differences between clouds. (**C**) Detail of the map of positive and negative differences and (**D**) its real color 3D realistic view, from which it is apparent that the differences correspond with remainders of vegetation.

The resulting cloud contains 597 million points, the average absolute distance of the clouds is 0.04 m, and the standard deviation is 0.087 m. A 0.001 m shift was determined for the Z-axis component and the standard deviation was 0.084 m. Figure 12 shows the whole captured area (Figure 12A), its analyzed part (talus cone, Figure 12B), a detail from a location where significant height deviations have been identified on the surface of the monitored area (Figure 12C), and the 3D representation of the same part of the terrain in real colors. It is obvious that the higher deviations correspond with the areas of low vegetation (strong red color on the differences point cloud corresponds with the area of dark green dwarf pines on the 3D realistic view), while dark blue spots in Figure 12C show deviations in areas between stones where the compared methods work differently. As SfM reconstructs the area using facets and has trouble reconstructing areas of deep shadow between stones while 3D scanning directly measures distances, under such specific terrain conditions, differences between the two models may arise not due to inaccuracy, but due to the difference in the calculation (measurement) principle.

After filtering these points out from the cloud based on the size of the deviation, i.e., 0.20 m, 537 million points remained in the cloud. The reduced point cloud showed a mean absolute deviation of 0.028 m and a standard deviation at an absolute distance of 0.029 m. A mean difference of −0.008 m is practically negligible in the altitude component, and the standard deviation is 0.032 m (see Table 5).

Table 5. Comparison of the point clouds—differences in the z axis.

Differences between Methods	Mean Difference (m)	Standard Deviation (m)	Mean Absolute Difference (m)	Maximum Negative Difference (m)	Maximum Positive Difference (m)
TLS vs. SfM	0.001	0.084	0.046	−3.620	4.993
TLS vs. SfM—filtered	−0.008	0.032	0.022	−0.200*	0.200*

Another representation of the errors present in the unfiltered cloud is shown in Figure 13. It clearly shows a cluster of outliers showing SfM overestimating the terrain by approx. 0.4 m, which exactly corresponds with the height of vegetation. Therefore, we can conclude that with the exception of vegetation-covered areas, the terrain model obtained from low-cost UAS photogrammetry achieves qualitative (precision) parameters comparable to those obtained by terrestrial laser scanning and is thus suitable as a basis for systematic monitoring that will form a basis for identification of surface changes at the centimeter level.

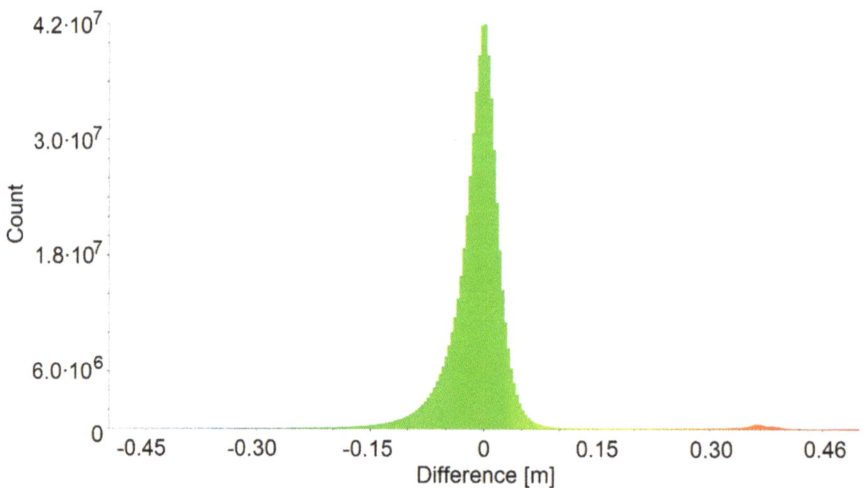

Figure 13. A histogram characterizing the distribution of the data differences in the unfiltered cloud showing a red group of outliers caused by low vegetation.

It must be also pointed out that our method directly compares point clouds rather than creating digital terrain models, creation of which requires a substantial aggregation of points and thus introduces errors. The method used, comparing each point from one point cloud against an area formed from an irregular triangular network based on nine nearest points from the other cloud, provides in our opinion the best possible accuracy estimation.

For monitoring purposes, our method comparing SfM models from different time points will therefore provide a clear record of vertical changes in the talus cones. A geodetic network with a sub-centimeter accuracy has been established on the site, which will allow easy re-setting of ground control points and repeated monitorings will therefore be much easier and less time-consuming.

The presented results also show the importance of removing outliers when comparing data acquired using different methods. The vegetation on the one hand and deep shadow areas combined with obstacles in the way of TLS beam on the other hand resulted in differences that were in some instances very high. After removing these outliers, however, the accuracy parameters detailed in Table 5 (and thus the agreement between methods) improved considerably (where values 0.200 marked by * are given by the filter limit values). Nevertheless, for the purposes of monitoring, even unfiltered clouds would provide valuable information as it is reasonable to assume that either the location of the vegetation would remain unchanged (hence there would be no or only minuscule difference when using the same method to capture it) or, if the entire talus cone moves, the vegetation would be either buried under the rubble or moved to a new location where it would be recorded as a change from the last monitoring record. In all instances, valuable information about the mass movement would be recorded and differences that were observed in our study between two methods based on different principles (SfM and TLS) would not play a major role for real monitoring.

5.2. Efficiency and Technical Demands of Selected Geodetic Methods

In the past 10 years, terrestrial laser scanning has found a solid place in various areas of geodesy, where the object of interest—the surface of objects or terrain—is documented in detail. When using TLS, the process of data retrieval consists of:

- Terrain reconnaissance and point field creation;
- Preparation of the instrument on the standpoint and subsequent scanning process;
- Processing of the measurements.

The time period for individual steps lasts for tens of minutes to hours, while the most demanding part is processing field measurements into the final model. In our case, due to the character of the terrain, the whole field data collection representing the transport of the scanner to the site and its movement on the rubble cone was very demanding both technically and physically. The terrestrial laser scanner is one of the most expensive surveying devices. Its weight is approx. 28 kg including the container, and the price is about €70,000. While carrying the device between the standpoints, any slip or fall of the person carrying it or the surveyor can mean damage or destruction of the instrument. Due to complicated transportation, mountain carrying, and manipulation with the scanner in difficult terrain, performing repeated measurements is difficult at best. This, along with the terrain ruggedness, resulting in obstruction of the terrain by terrain features closer to the scanner, speaks against the use of this technology in mountain conditions.

In this field, the opposite of using TLS is the use of cheap UAS carriers. Transport of this technology over the mountain terrain is incomparably easier. The cost of a "cheap" UAS is around €1500 and the weight is about 3 kg, which is a fraction of the price and weight of the TLS. Overall, from the point of view of efficiency and technical difficulty, repeated monitoring by using low-cost UAS photogrammetry is a much preferable method for collecting data in mountain conditions.

5.3. Suitability of Used Geodetic Methods to Monitor Dynamics of the Development of the Georelief

For real use of selected geodetic methods and instruments, based on the achieved results, it should be pointed out that despite the expected decreased accuracy of low-cost UAS photogrammetry, UAS imaging and subsequent SfM image processing are a significantly better choice than TLS. In addition to being an easier and less time-consuming measurement, it also offers a complex coverage of the monitored area without "holes" in the point cloud, and includes very good color information to allow easy interpretation of the detected shifts. In terms of laboriousness, there is a big difference between methods—drone imaging, including stabilization of control points, took about 3 h. Apart from the stabilization of control points mentioned above, it is not necessary to move in the monitored area, which is extremely advantageous due to its surface. In contrast, 3D terrestrial scanning took two working days, and the movement in the stone field with the measuring technique was physically very demanding and lengthy, which makes UAS a significantly more economical solution. Speaking of the economical side of model acquisition, the equipment for low-cost drone imaging is also significantly cheaper. While the hardware and software costs were below €5000 in the configuration used, it amounts to approx. €100,000 in the case of 3D scanning (both without geodetic instruments for geodetic surveying, network and control points).

Besides, when using TLS scanning, the terrain morphology and the need to scan even from low attitudes causes discontinuous terrain capture due to many obstructions from nearby large boulders when creating the final point cloud.

Another advantage of the UAS photogrammetry is the production of high-resolution ortho-photos in addition to point clouds, which facilitates visual identification of changes.

In our study, we did not attempt to record any temporal changes in the mass movement but aimed at developing methods for future recording of such changes. Although the accuracy of both tested methods is sufficient for intended monitoring of changes, it must be taken into account that due to differences in measurement and data processing, the model can produce apparent changes that are not substantiated. Each of these technologies is, on its own, sufficiently accurate given the accuracy, but according to efficiency and overall suitability of the methods for demanding high-altitude conditions and the appearance of the other parameters described above, it is unambiguously recommended to use the more suitable drone imaging method, which is fast, sufficiently accurate, and environmentally friendly. It would be interesting to try 3D scanning from the drone for this purpose, but this would be associated with significantly higher input costs and again introduce a risk of damaging very expensive equipment. Besides, airborne 3D scanners are substantially heavier than a simple camera and would require a drone with a higher load capacity, which would be associated with higher noise generation. Operating such a noisy drone would be, especially in the tranquil environment of the national park, highly undesirable.

6. Conclusions

In our study, we have shown that in difficult mountain conditions, UAS photogrammetry is capable of providing models of comparable quality with those acquired by TLS for monitoring movement of the talus cones. Taking into account the ease of data acquisition, equipment costs, and risk of damage of the expensive TLS equipment, UAS photogrammetry appears to be an ideal method for repeated monitoring of geological phenomenons in a very rugged mountainous terrain due to its accuracy, low cost, ease of application, and efficiency, plus it is technically unassuming.

Our method used a direct comparison of point clouds rather than creating digital terrain models that inherently require some level of approximation, thus introducing error. Besides, although we demonstrated the level of improvement of the agreement between TLS and SfM by filtering out outliers (mostly caused by remnants of vegetation in the talus cone or by principal differences between the way both methods operate), real monitoring would only use the SfM method and as such, recording the changes (i.e., mass movement) would in all likelihood work very well even without this filtering out of outliers as both above reasons for the error would be prevented.

Author Contributions: Conceptualization, R.U. and M.Š.; Methodology, P.B., Ľ.K., S.J., I.Ď.; Software, M.Š.; Validation, R.U., P.B. and Ľ.K.; Formal Analysis, M.K. and S.S.; Investigation, R.U., M.Š., P.B., Ľ.K. and S.J.; Resources, M.Š. and P.B.; Data Curation, P.B., M.P. and I.Ď.; Writing—Original Draft Preparation, R.U. and M.Š.;

Writing—Review and Editing, R.U. and M.Š.; Visualization, P.B.; Supervision, P.B.; Project Administration, P.B.; Funding Acquisition, M.K. and S.S.

Funding: This research was funded by: Grant Agency of CTU in Prague—grant number SGS19/047/OHK1/1T/11, Scientific Grant Agency of the Slovak Republic (VEGA – MŠVVaŠ SR—grant number 1/0844/18 and supported by project SKHU/1601/4.1/187.

Acknowledgments: This study would not have been possible without the support of the National Park—TANAP.

Conflicts of Interest: The authors declare no conflict of interest. The funders had no role in the design of the study; in the collection, analyses, or interpretation of data; in the writing of the manuscript, or in the decision to publish the results.

References

1. Braun, J.; Hánek, P. Geodetic Monitoring Methods of Landslide-Prone Regions—Application to Rabenov. *Acta Universitatis Carolinae Geographica* **2014**, *49*, 5–19. [CrossRef]
2. Barbarella, M.; Fiani, M. Monitoring of large landslides by terrestrial laserscannig techniques: Field data collection and processing. *Eur. J. Remote Sens.* **2013**, *46*, 126–151. [CrossRef]
3. Bitelli, G.; Dubbini, M.; Zanutta, A. Terrestrial Laserscannig and Digital Photogrammetry Tecniques to Monitor Landslide Bodies. In Proceedings of the XX-th ISPRS Congress: Commission V, Istanbul, Turkey, 12–24 July 2004; pp. 246–251.
4. Derrien, A.; Villeneuve, N.; Peltier, A.; Beauducel, F. Retrieving 65 years of volcano summit deformation from multitemporal structure from motion: The case of Piton de la Fournaise (La Réunion Island). *Geophys. Res. Lett.* **2015**, *42*, 6959–6966. [CrossRef]
5. Jovančević, S.D.; Peranić, J.; Ružić, I.; Arbanas, Ž. Analysis of a historical landslide in the Rječina River Valley, Croatia. *Geoenviron. Disasters* **2016**. [CrossRef]
6. Rossi, G.; Tanteri, L.; Tofani, V.; Vannocci, P. Multitemporal UAV surveys for landslide mapping and characterization. *Landslides* **2018**, *15*, 1045. [CrossRef]
7. Ridolfi, E.; Buffi, G.; Venturi, S.; Manciola, P. accuracy analysis of a dam model from drone surveys. *Sensors* **2017**, *17*, 1777. [CrossRef]
8. Buffi, G.; Manciola, P.; Grassi, S.; Barberini, M.; Gambi, A. Survey of the Ridracoli Dam: UAV–based photogrammetry and traditional topographic techniques in the inspection of vertical structures. *Geomat. Nat. Hazards Risk* **2017**, 1562–1579. [CrossRef]
9. Duró, G.; Crosato, A.; Kleinhans, M.G.; Uijttewaal, W.S.J. Bank erosion processes measured with UAV-SfM along complex banklines of a straight mid-sized river reach. *Earth Surf. Dyn.* **2018**, *6*, 933–953. [CrossRef]
10. Peppa, M.V.; Mills, J.P.; Moore, P.; Miller, P.E.; Chambers, J.E. Brief communication: Landslide motion from cross correlation of UAV-derived morphological attributes. *Nat. Hazards Earth Syst. Sci.* **2017**, *17*, 2143–2150. [CrossRef]
11. Salvini, R.; Mastrorocco, G.; Esposito, G.; Di Bartolo, S.; Coggan, J.; Vanneschi, C. Use of a remotely piloted aircraft system for hazard assessment in a rocky mining area (Lucca, Italy). *Nat. Hazards Earth Syst. Sci.* **2018**, *18*, 287–302. [CrossRef]
12. Kovacevic, M.S.; Car, M.; Bacic, M.; Stipanovic, I.; Gavin, K.; Noren-Cosgriff, K.; Kaynia, A. *Report on the Use of Remote Monitoring for Slope Stability Assessments*; H2020-MG 2014-2015; Innovations and Networks Executive Agency: Zagreb, Croatia, 2015.
13. Kaufmann, V.; Seier, G.; Sulzer, W.; Wecht, M.; Liu, Q.; Lauk, G.; Maurer, M. Rock Glacier Monitoring Using Aerial Photographs: Conventional vs. UAV-Based Mapping—A Comparative Study. *Int. Arch. Photogram. Remote Sens. Spatial Inf. Sci.* **2018**, *XLII-1*, 239–246. [CrossRef]
14. Vivero, S.; Lambiel, C.H. Monitoring the crisis of a rock glacier with repeated UAV surveys. *Geogr. Helv.* **2019**, *74*, 59–69. [CrossRef]
15. Kohút, M.; Janák, M. Granitoids of the Tatra Mts., Western Carpathians: field relations and petrogenetic implications. *Geol. Carpath.* **1994**, *45*, 301–311.
16. Petrík, I.; Nabelek, P.I.; Janák, M.; Plašienka, D. Conditions of Formation and Crystallization Kinetics of Highly Oxidized Pseudotachylytes from the High Tatras (Slovakia). *J. Petrol.* **2003**, *44*, 901–927. [CrossRef]

17. Lukniš, M. The course of the last glaciation of the Western Carpathians in the relation to the Alps, to the glaciation of northern Europe, and todivison of central European Würm into periods. *Geografický Casopis* **1964**, *16*, 127–142.

18. Klimaszewski, M. *Relief of the Polish Tatra Mountains*; Panstwowe Wydawnictwo Naukowe: Warszawa, Poland, 1988.

19. Lindner, L.; Dzierzek, J.; Marciniak, B.; Nitychoruk, J. Outline of Quaternary glaciations in the Tatra Mts.: their development age and limits. *Geol. Q.* **2003**, *47*, 269–280.

20. Zasadni, J.; Klapyta, P. The Tatra Mountains during the Last Glacial Maximum. *J. Maps* **2014**, *10*, 440–456. [CrossRef]

21. Kotarba, A.; Strömquist, L. Transport, sorting and deposition processes in the Polish Tatra Mountains. *Geogr. Ann.* **1984**, *66*, 285–294. [CrossRef]

22. Kotarba, A.; Kaszowski, L.; Krzemień, K. *High-Mountain Denudational System in the Polish Tatra Mountains*; Ossolineum: Wroclaw, Poland, 1987; pp. 1–106.

23. Sass, O.; Krautblatter, M. Debris-flow-dominated and rockfall-dominated scree slopes: genetic models derived from GPR measurements. *Geomorphology* **2007**, *86*, 176–192. [CrossRef]

24. Fort, M.; Cossart, E.; Deline, P.; Dzikowski, M.; Nicoud, G.; Ravanel, L.; Schoeneich, P.; Wassmer, P. Geomorphic impacts of large and rapid mass movements: a review. *Geomorpholgie* **2009**, *1*, 47–63. [CrossRef]

25. Krautblatter, M.; Moser, M.; Schrott, L.; Wolf, J.; Morche, D. Significance of rockfall magnitude and carbonate dissolution for rock slope erosion and geomorphic work on Alpine limestone cliffs (Reintal, German Alps). *Geomorphology* **2012**, *167–168*, 21–34. [CrossRef]

26. Lukniš, M. *Reliéf Vysokých Tatier a Ich Predpolia*; Vydavateľstvo SAV: Bratislava, Slovak, 1973; p. 375.

27. Zimmerman, M.; Haeberl, I.W. Climate change and debris flow activity in the highmountain areas—A case study in the Swiss Alps. *Catena Suppl.* **1990**, *22*, 59–72.

28. Kotarba, A. Formation of high mountain talus slope related to debris-flow activity in the High Tatra Mountains. *Permafr. Periglac. Process.* **1997**, *8*, 191–204. [CrossRef]

29. Anderson, R.S. Modeling the tor-dotted crests, bedrock edges, and parabolic profiles of high alpine surfaces of the Wind River Range, Wyoming. *Geomorphology* **2002**, *46*, 35–58. [CrossRef]

30. Melo, M.; Lapin, M.; Pecho, J.; Kružicová, H. Climate Trends in the Slovak Part of the Carpathians. In *The Carpathians: Integrating Nature and Society Towards Sustainability*; Springer: Berlin/Heidelberg, Germany, 2013; ISBN 978-3-642-12724-3. [CrossRef]

31. Kotarba, A. Geomorphic Processes and Vegetation Pattern Changes. Case Study in the Zelené Pleso Valley, High Tatra, Slovakia. *Studia Geomorphol. Carpatho Balcanica* **2005**, *39*, 39–48.

32. Van Steijn, H. Debris-flow magnitude-frequency relationships for mountainous regions of central and northwest Europe. *Geomorphology* **1996**, *15*, 259–273. [CrossRef]

33. Niedźwiedz, T. Extreme Precipitation Events on the Northern Side of the Tatra Mountains. *Geogr. Polonica.* **2003**, *76*, 15–23.

34. Rusnák, M.; Sládek, J.; Kidova, A.; Lehotský, M. Template for high-resolution river landscape mapping using UAV technology. *Measurement* **2018**, *115*. [CrossRef]

35. Software CloudCompare. Available online: www.cloudcompare.org (accessed on 10 April 2019).

International Journal of
Geo-Information

Article

Diachronic UAV Photogrammetry of a Sandy Beach in Brittany (France) for a Long-Term Coastal Observatory

Marion Jaud [1,*], Christophe Delacourt [2], Nicolas Le Dantec [2,3], Pascal Allemand [4], Jérôme Ammann [2], Philippe Grandjean [4], Henri Nouaille [2], Christophe Prunier [2], Véronique Cuq [5], Emmanuel Augereau [6], Lucie Cocquempot [2] and France Floc'h [2]

[1] IUEM—UMS 3113, Université de Bretagne Occidentale, IUEM, CNRS, Technopôle Brest-Iroise, Rue Dumont d'Urville, F-29280 Plouzané, France

[2] Laboratoire Géosciences Océans-UMR 6538, Université de Bretagne Occidentale, IUEM, CNRS, Technopôle Brest-Iroise, Rue Dumont d'Urville, F-29280 Plouzané, France; christophe.delacourt@univ-brest.fr (C.D.); nicolas.ledantec@univ-brest.fr (N.L.D.); jerome.ammann@univ-brest.fr (J.A.); henri.nouaille@gmail.com (H.N.); christophe.prunier@univ-brest.fr (C.P.); lucie.cocquempot@univ-brest.fr (L.C.); france.floch@univ-brest.fr (F.F.)

[3] Cerema, Direction Eau Mer et Fleuves, F-60280 Margny-lès-Compiègne, France

[4] Laboratoire de Géologie de Lyon: Terre, Planètes, Environnement—UMR 5276, Université de Lyon, Université Claude Bernard Lyon 1, ENS Lyon, CNRS, F-69622 Villeurbanne, France; pascal.allemand@univ-lyon1.fr (P.A.); philippe.grandjean@univ-lyon1.fr (P.G.)

[5] LETG—UMR 6554, Université de Bretagne Occidentale, IUEM, CNRS, Technopôle Brest-Iroise, rue Dumont d'Urville, F-29280 Plouzané, France; Veronique.Cuq@univ-brest.fr

[6] OSU de La Réunion, Université de La Réunion, CS 92003, F-97744 Saint-Denis, France; emmanuel.augereau@univ-brest.fr

[*] Correspondence: marion.jaud@univ-brest.fr; Tel.: 33-298-498-891

Received: 7 May 2019; Accepted: 5 June 2019; Published: 7 June 2019

Abstract: In the dual context of coastal hazard intensification and the growing number of stakes exposed to these hazards, coastal observatories are in demand to provide a structured framework dedicated to long-term monitoring. This article describes the drone-based photogrammetry monitoring performed since 2006 on Porsmilin Beach (Brittany, France) in the framework of the DYNALIT (Littoral and Coastline Dynamics) observatory, focusing on data quality and the consistency of long-term time series under the influence of multiple technological evolutions: Unmanned Aerial Vehicles (UAV) platforms with the arrival of electric multirotor drones, processing tools with the development of structure-from-motion (SfM) photogrammetry, and operational modes of survey. A study case is presented to show the potential of UAV monitoring to study storm impacts and beach resilience. The relevance of high-accuracy monitoring is also highlighted. With the current method, an accuracy of 3 cm can be achieved on the digital elevation model (DEM) and the orthophotograph. The question of the representativity and frequency of DEM time points is raised.

Keywords: UAV; SfM photogrammetry; coastal observatory; beach monitoring

---

1. Introduction

Coastal zones are exposed to both marine and terrestrial processes, as well as anthropogenic impacts. As a fragile interface subject to highly dissipative energetic processes, coastal zones can undergo fast morphological changes. Furthermore, coastal zones are highly exposed to several natural risks: submersion, erosion, pollution, and so forth. These coastal risks are increasing due to the combination of (i) the intensification of the natural hazards driven by a relative shortage of sediment stocks and global climate change, and (ii) the growing number of stakes exposed to coastal hazards.

Shoreline change is the result of both natural and anthropogenic factors and occurs over a wide range of spatial and temporal scales. Coastal zone management, therefore, requires adequate monitoring to unravel this multiscale evolution [1,2]. It is essential to improve the knowledge and tools needed to understand and quantify the physical evolution of the coastal environment and the processes behind these morphological changes [3–7]. In this context, coastal observatories offer a structured framework for long-term monitoring in a coordinated and systematic manner [8–10]. Such observatories are valuable both for scientific research and to provide usable information for management decisions and coastal engineering policies.

DYNALIT, created in 2014, is the French National Observation Service of CNRS/INSU (French National Institute of the Universe Sciences) dedicated to long-term observation of the coastal dynamic (https://www.dynalit.fr/). The DYNALIT network involves 22 universities and regroups 36 observation sites distributed over the French metropolitan and ultramarine coastlines, covering a wide range of coastal environments. Among these coastal environments, sandy beaches are particularly challenging to monitor, as they feature rapid and sometimes high-amplitude morphological changes and continually adjust to numerous processes driving morphological changes.

This study focused on the site of Porsmilin, a sandy embayed beach in Brittany, which is a DYNALIT study site. This beach has been monitored since 2000 with monthly surveys of topographical cross-shore profiles measured by total station or real-time kinematic differential GPS (RTK DGPS) [11]. Two-dimensional cross-shore profiles have been followed by digital elevation models (DEMs) computed from Unmanned Aerial Vehicles (UAV) photographs since 2006 or from terrestrial laser scanner (TLS) point clouds collected since 2009 [12,13]. A video imagery system has also been set up, continuously collecting one image per hour (and up to four images per second) since 2014. Intensive field campaigns of hydrodynamic measurements were also carried out in 2014 and 2016 [14,15]. The DYNALIT observatory is compatible with the on-going evolution of methodologies, as well as the diversity of observation approaches.

The potential of the UAV photogrammetric technique for monitoring geomorphological changes and, particularly, the evolution of sandy beaches has been demonstrated for several years [16,17]. Nevertheless, with technological improvements in platforms, sensors, and processing algorithms occurring at a fast pace, the method has significantly evolved over the years. This article focuses more particularly on long-term UAV photogrammetry, including how the methodological evolutions have been incorporated into the monitoring protocol and how these operational changes affect the quality of the monitoring.

2. Study Area

Porsmilin Beach (Figure 1) is located at the entrance of the Bay of Brest in the Iroise Sea in Brittany, France, a macrotidal zone with semidiurnal and symmetrical tides and a mean spring tidal range of 5.7 m. The Iroise Sea is a highly energetic environment, with annual and decadal significant wave heights of, respectively, 11.3 and 14.5 m in 110 m of water depth to the west of Brittany [18]. On the beach, the mean annual wave height is 0.5 m, whereas storm waves are up to 2 m high [18]. Inland and to the north, Porsmilin Beach is backed by a brackish-water marsh that no longer communicates with the sea. To the east and west, it is flanked by orthogneiss cliffs of about 15 m in height and bounded by headlands extending offshore into rocky reefs that only allow incident waves from the southwest and act as obstacles for longshore sand transport [14].

The intertidal zone is around 200 m wide in the longshore direction. The median sediment grain size (D50) is 320 μm [18], with significant cross-shore variability, including coarser sediments (D50 ≈ 700 μm) around the crest of intertidal bars and the existence of shingles on the upper beach which become uncovered during episodes of energetic hydrodynamic forcing [14].

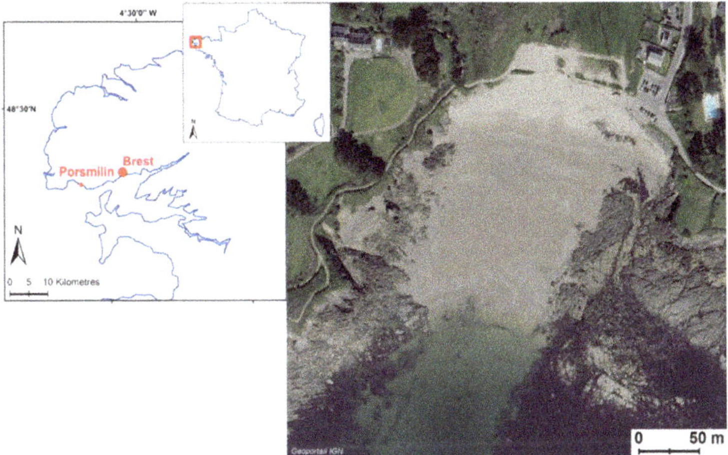

Figure 1. Location and orthophotograph of Porsmilin Beach (Geoportail IGN©—2015).

3. Evolution of Technical Equipment and Methods

3.1. Evolution of the UAV Platforms

The first UAV platform used above Porsmilin Beach was a Pixy® drone (Figure 2, Table 1), a model developed by the IRD (French Institute for Research and Development) and PHILAE concept® society [19]. It was a small radio-controlled paramotor, composed of a 3.8 m² tubular fabric wing maintained by suspension lines. A tripod structure, suspended below the wing, supported a combustion engine of 25 cm³ and the sensors, including the camera (Fuji S2 Pro—50 mm). The platform weighed 7 kg and had a maximum payload of 4 kg. The flight was performed between 15 and 35 km/h. This model was used to perform surveys above Porsmilin Beach between 2006 and 2008. One of the main drawbacks of this system was the requirement of a wide area for takeoff and landing.

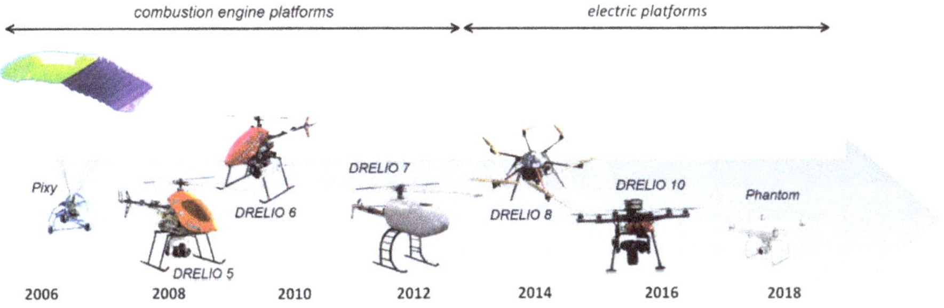

Figure 2. Evolution of the Unmanned Aerial Vehicles (UAV) platforms used for Porsmilin Beach monitoring.

From 2007 to 2012, three remote-controlled combustion engine helicopters (Figure 2, Table 1) were used for Porsmilin Beach monitoring. These UAV platforms, called DRELIO (DRone for Environmental and Littoral Observations), had been specifically adapted to coastal environments [16]: protection against corrosive salted air, ability to endure wind gusts, and so forth. These helicopters weighed around 10 kg with an additional payload of 5–8 kg. The payload was modular, but for photogrammetric purposes, DRELIO helicopters were equipped with a commercial reflex camera (Nikon D200 at 10 Mpix and Nikon D700 at 12 Mpix), with a 35 mm focal length and parametrized in intervalometer mode.

The autopilot was connected to an atmospheric pressure sensor, an inertial sensor, a geomagnetic direction sensor, and a GPS. DRELIO helicopters were capable of performing fully automatic takeoff and landing and following hovering flight plans [16]. The combustion engine enabled a flight autonomy of 45–50 min. Significant efforts had been made to limit vibrations, including reducing the rotation speed of blades or improving the stability of the camera gimbal. For DRELIO 7, the design was more industrialized and the structure was conceived with shock absorbers to isolate the payload from the motor vibrations.

Table 1. Main characteristics of the different UAV platforms used for Porsmilin Beach monitoring.

Drone	Operating Period	Characteristics			Camera	Range of Mean Accuracy
		Weight	Payload	Autonomy		
Pixy	2006–2008	7 kg	4 kg	~1 h	Fuji S2 Pro, 50 mm, 6 Mpix	10–20 cm
DRELIO 5	2007–2010	11 kg	5 kg	45 min	Nikon D200, 35 mm, 10 Mpix	5–20 cm
DRELIO 6	2010	9.5 kg	5 kg	45 min	Nikon D200, 35 mm, 10 Mpix	5–20 cm
DRELIO 7	2012	15 kg	8 kg	50 min	Nikon D700, 35 mm, 12 Mpix	5–10 cm
DRELIO 8	2013–2014	<4 kg	1.5 kg	15 min	Nikon D700, 35 mm, 12 Mpix	1–5 cm
DRELIO 10	2015–2019	<4 kg	1.5 kg	15 min	Nikon D800, 35 mm, 36 Mpix	1–2 cm
Phantom Pro	2018–2019	1.4 kg	0.3 kg	25 min	FC6310, 8.8 mm, 20 Mpix	1–3 cm

Since 2013, the DRELIO platforms that have been used in Porsmilin Beach are electrical multirotors (Figure 2, Table 1): hexacopters (MikroKopter® and DroneSys DS6®) or quadcopters (DJI Phantom®). Both hexacopters are almost the same, weighing less than 4 kg and being able to handle a payload of 1.5 kg. Their actual flight autonomy is about 15 min. The flight control is performed by the DJI® software iOSD. The payload is modular and, as before, for photogrammetric purposes, a commercial reflex camera (Nikon D700 at 12 Mpix or Nikon D800 at 36 Mpix) with a 35 mm focal length is used [20]. The Phantom 4 Pro drone weighs less than 1.5 kg (including batteries and camera). This inexpensive, ready-to-use system is also a highly integrated solution, limiting the possibilities of tuning the drone or the sensor. Now, DJI® proposes a Phantom 4 equipped with RTK positioning, but this model has not been tested at Porsmilin Beach yet.

Compared with the previous models, the multirotor platforms have shorter flight autonomy but are much easier to control and easier to transport. Furthermore, they induce fewer vibrations and allow more stable flights. With the first drone platforms, some parts of the study area were not covered by the photographs, resulting in incomplete DEMs and orthophotographs. The vibrations also caused blurring effects on certain photographs (Figure 3), which could make tie-point detection challenging. Further, these vibrations generated less stability of the inner orientation of the camera, which made correcting the associated systematic errors more difficult.

Autopilots have also evolved. The Pixy® drone was not equipped with an autopilot and was only remotely piloted. On the helicopter platforms, the system was designed for aerobatic aircraft models. The operator had to parametrize the speed of the drone and the camera intervalometer as a function of the flight height and the desired overlap, with the problem of stability loss at very low speed. Between 2006 and 2014, the flight plan was parametrized to achieve along-track and across-track overlap of around 60%. Since 2015, the along-track overlap has been increased to 80%. With the Phantom drone, as the autopilot is designed for photogrammetric applications, the overlap is directly parametrized in the flight plan.

Figure 3. Portion of an orthoimage illustrating the variation in photograph quality during a flight performed in June 2008 with the Pixy® drone.

3.2. Evolution of the Processing Tools

Originally, data processing was performed using codes written with Matlab® software. The inner orientation and lens distortion parameters of the camera were predetermined from a series of photographs collected before the survey above a grid pattern of georeferenced targets in such a way that the targets were visible in different parts of the images (Figure 4). The stereorestitution algorithms were based on classical photogrammetric techniques [16].

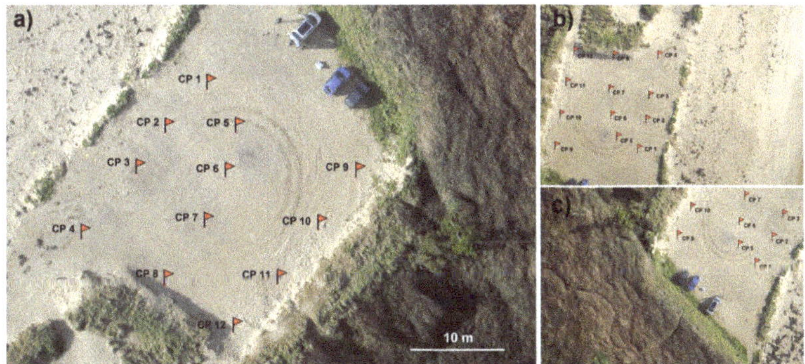

Figure 4. Examples of photographs collected by DRELIO 5 in February 2009 above the grid pattern of targets for camera calibration.

During the last decade, stereophotogrammetric processing has largely evolved with the development of the structure-from-motion (SfM) technique. This method, based on computer vision concepts, solves automatically and simultaneously the camera self-calibration and scene geometry using image matching and bundle adjustment. SfM photogrammetry is therefore less constraining for the camera calibration, which simplifies the in situ survey protocol. At the same time, the generalization of the scale-invariant feature transform (SIFT) algorithm [21] has changed the approach of tie-point detection, enabling scale- and rotation-invariant image matching.

In the context of these methodological developments, photogrammetric software tools have been popularized, accompanying a growing use of photogrammetry outputs for many applications. Many open-source and commercial solutions are now proposed. On the Porsmilin Beach study site, we mainly used Agisoft PhotoScan Pro® software, and occasionally MicMac®, the free open-source software produced by IGN© (the French National Institute of Geographic and Forest Information). For consistency, all the former surveys of the observatory datasets were reprocessed using PhotoScan®, following the same protocol.

3.3. Evolution of the Survey Operational Mode

The DYNALIT Observation Service encompasses a great diversity of sites (in view of their type, their size, the implied hydrodynamic processes, etc.). Various survey methods are therefore implemented depending on the observation site. For Porsmilin Beach, the evolution in equipment and processing tools has induced an evolution of the survey operational mode.

Regardless of the survey platform, acquisition protocol, and processing tools, ground control points (GCPs) are essential in the photogrammetric process to attain centimetric accuracy [16,22–24]. Therefore, this issue has been received much attention. As the beach is permanently changing, it is not possible to use fixed GCPs. Instead, different kinds of removable targets have been tested as GCPs (Figure 5). These targets have to be highly visible on the beach with various illumination conditions. They also need to be sand and water resistant, easy to transport, and low cost. For example, the light green and orange square plastic targets (33 × 33 cm) used for the first surveys of Porsmilin Beach (Figure 5a,b) were sometimes difficult to detect in the photographs. Currently, GCPs consist of red plastic plates that are 30 cm in diameter (Figure 5c), which can be secured with pegs in case of wind or wave surge. Monochromic targets have been chosen because they appear to be easier to detect on the upper part of the beach, which is highly reflective in sunny weather. Furthermore, the red color rarely occurs in natural environments and strongly contrasts with pale or bright sand.

Figure 5. Examples of the different types of targets used as ground control points (GCPs): (**a,b**) light green and orange square plastic targets with varying illumination conditions (photos respectively taken in 2006 and 2008); (**c**) 30 cm red plastic plate currently used (photo taken in 2015).

The deployment and positioning of the targets is the most time-consuming step of the survey process. The flexibility of SfM photogrammetry, particularly with the benefits of self-calibration, compared with classical photogrammetry allows using a limited number of GCPs. However, it is still critical to optimize the geometry of the GCP network in order to avoid geometric distortion [23–25].

The method used to accurately measure GCP location has also changed. Since 2006, GCPs have been surveyed by RTK DGPS most of the time, and occasionally with a total station. Originally, the GPS base antenna was situated 900 m away from the beach. As the measurement accuracy decreases with the distance between base antenna and rover antenna, a new topographic marker for the GPS base station position was set up near Porsmilin Beach in 2010. To ensure continuity and redundancy in the quality of GCP measurements, three control points, also marked by topographic nails, are located close to the beach. In 2010, the reference coordinate system was switched from NGF–Lambert 1 to RGF 93–Lambert 93, which is currently the official coordinate system in France. All the first GPS point files have since been converted to RGF 93–Lambert 93 using the reference IGN® Circé converter tool. When the former surveys of the observatory datasets were reprocessed using PhotoScan® to ensure consistency, the converted GPS point files (in RGF 93–Lambert 93) were used for GCP locations in data reprocessing.

4. Results and Discussion

4.1. Data Quality

Changes to the survey equipment, protocols, and data processing have been made over time with the objective of optimizing the consistency of the dataset series. When possible, in particular for processing steps, the earlier datasets were reprocessed using the same workflow. Still, the quality of the oldest datasets is lower than the quality currently attained. The improvement of quality with time is due to various factors, such as: (i) the technological evolutions of the platform (allowing more flexibility in flight plan designing and more stability in flight), (ii) the technological evolutions of the sensors (allowing higher spatial resolution), (iii) higher accuracy in GCP measurement, and (iv) improvements in the survey protocol (for example, an optimization of the GCP network). These evolutions have driven the monitoring strategy of the observatory for Porsmilin Beach regarding the choice of survey instrumentation and methods. Until 2009, UAV photogrammetry was the only method available in the observatory to conduct very high resolution surface surveys over large areas. With the acquisition of TLS in 2009, UAV surveys became more scattered, as TLS surveys were, at that time, more accurate, easier to organize, and easier and quicker to process. Since 2014–2015, with the aforementioned evolutions, UAV photogrammetry is somewhat superseding TLS surveys in the data acquisition strategy of the observatory.

Ideally, the quality of the dataset is assessed using some georeferenced targets as control points rather than GCPs. In the dataset of June 2015, which is representative of current datasets, the root-mean-square error (RMSE) was lower than 3 cm (using 12 GCPs and calculating errors on 16 validation points). For several datasets where all the targets were used as GCPs because the GCP network was not optimal, the quality is difficult to evaluate. In such cases, the accuracy of the DEM and the orthophotograph is assessed by comparing the position of some remarkable points in the study area (e.g., rocks, boat ramp, or car park) with their position on a dataset with the highest accuracy (here, June 2015). However, this method only provides a first-order error assessment based on very few points located on the edge of the study area, where image overlapping is not as good as on the beach itself. Among the oldest datasets (from 2006 to 2008), the mean accuracy is about 10–15 cm, with errors sometimes up to 90 cm on certain parts of the DEM with poor photograph overlap or poor tie-point distribution.

4.2. Benefits of DEM Surveys

In the context of the coastal observatory, the DEM and the orthophotograph provide richer spatial information than cross-shore profiles. Indeed, they enable to capture cross-shore movements but also long-shore variations of the beach morphology, such as beach cusps (Figure 6). Moreover, they allow a multiscale approach, from the study of the whole beach to the morphodynamics of smaller-scale sedimentary structures. Above all, the computation of DEMs of differences (DoDs), corresponding to grids of changes in elevation that occurred between two acquisition dates, provides a diachronic evolution of the entire beach. From these DoDs, sedimentary budgets can be computed with better spatial representativity than from GPS profiles.

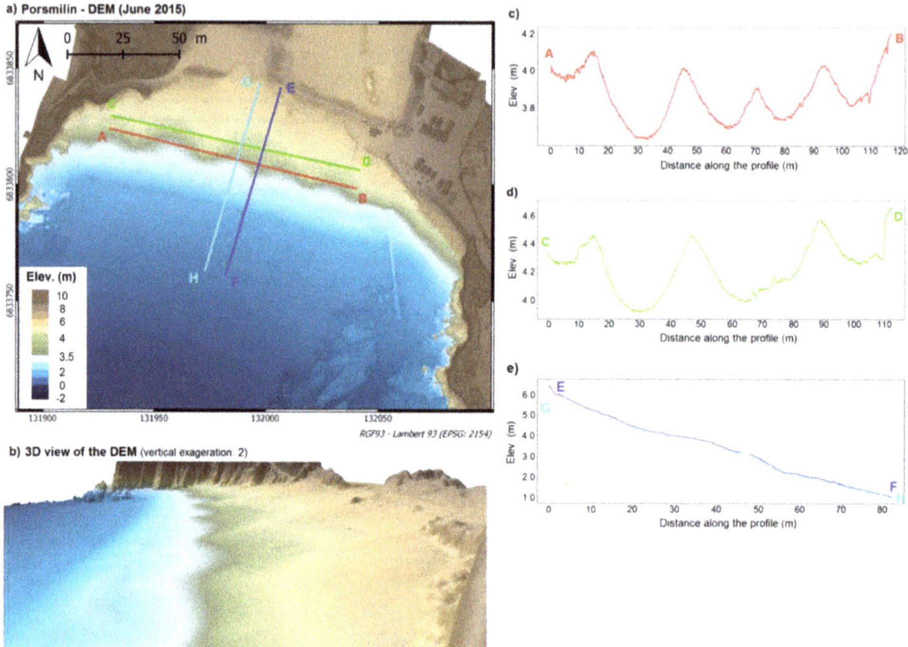

Figure 6. (**a**) Portion of the digital elevation model (DEM) computed from the photographs collected on the 16 June 2015. (**b**) 3D view of the DEM showing the beach cusps (vertical exageration: 2). (**c**,**d**) Alongshore beach vertical profiles showing the beach cusp morphology. (**e**) Cross-shore profiles showing differences in beach face gradient between embayments and horns.

4.3. Study of Beach Resilience Following a Major Storm

Figure 7 presents an example illustrating the relevance of UAV datasets for coastal observation and, particularly, studying beach resilience. During winter 2013–2014, successive storms hit Western Europe [26–28] causing severe erosion and submersion damage. A UAV dataset was collected in Porsmilin in February 2014, just after a series of storms: Petra (2014/02/04–05), Quimaira (2014/02/06–07), Ruth (2014/02/08), Tini (2014/02/12), and Ulla (2014/02/14). The morphology of the upper section of Porsmilin Beach was deeply altered, provoking a coastline retreat up to 18 m [29]. The orthophotograph and the DoD show this evolution, providing evidence of the water levels attained during storms and allowing computation of sediment budgets. The comparison of the orthophotographs of March 2012 (Figure 7a), February 2014 (Figure 7b), and June 2015 (Figure 7c) shows that the coastline position and the back-beach facilities were deeply and durably modified. Table 2 shows the pairwise differential sediment balance and normalized mean volume change evaluated between the three survey dates and over the computation zone (5360 m^2, represented in Figure 7d–f). Considering the spatial distribution of the sediment budget provided by the DoDs, it appears that the erosion mainly occurred on the upper shore, where the coastline had retreated. On the beach itself, the erosion that occurred during the storm was balanced by new sediment deposits, confirming the tendency of the beach to return to its equilibrium state.

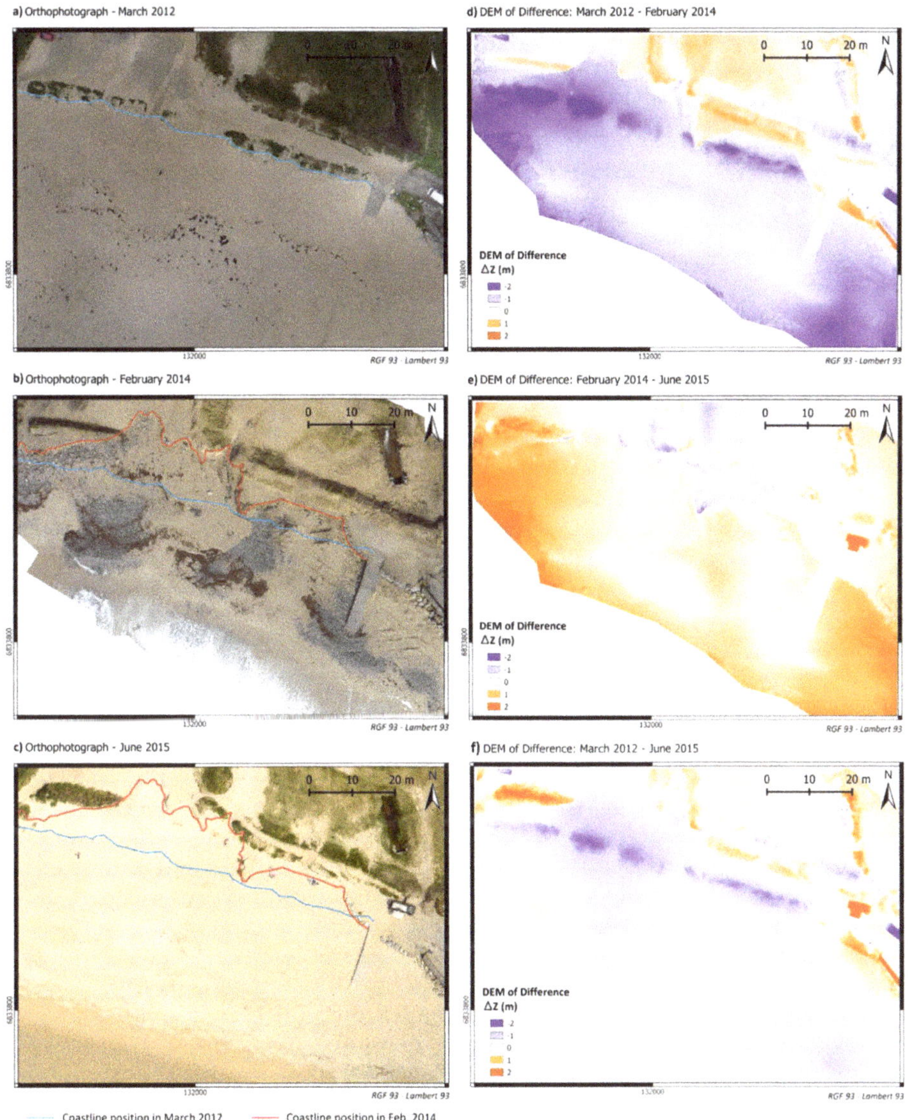

Figure 7. Use of orthophotographs and DoDs to study the impact of the storms (winter 2013–2014). Comparison of the orthophotographs in March 2012 (**a**), February 2014 (**b**), and June 2015 (**c**). DEM of differences computed over the periods from March 2012 to February 2014 (**d**), February 2014 to June 2015 (**e**), and March 2012 to June 2015 (**f**).

Table 2. Pairwise differential sediment balance and normalized mean volume change.

DEMs of Differences (DoDs)	Sediment Budget (m^3)	Normalized Mean Volume Change (m^3/m^2)
March 2012–Feb 2014	3005 m^3 ($\pm$ 536 m^3)	−0.56 m^3/m^2
Feb 2014–June 2015	2002 m^3 ($\pm$ 536 m^3)	+0.37 m^3/m^2

The DoD between March 2012 and March 2017 (Figure 8) enabled observation of the resilience of Porsmilin Beach over a longer period. Being both created from datasets of March, the DEMs considered in the computation of this DoD were chosen to minimize the impact of seasonal effects on this long-term sedimentary budget. The sediment balance was an erosion of -2650 m^3 ($\pm$ 1708 m^3) over an area of 17,083 m^2, which corresponded to a normalized mean erosion of -0.15 m^3/m^2 ($\pm$ 0.10 m^3/m^2).

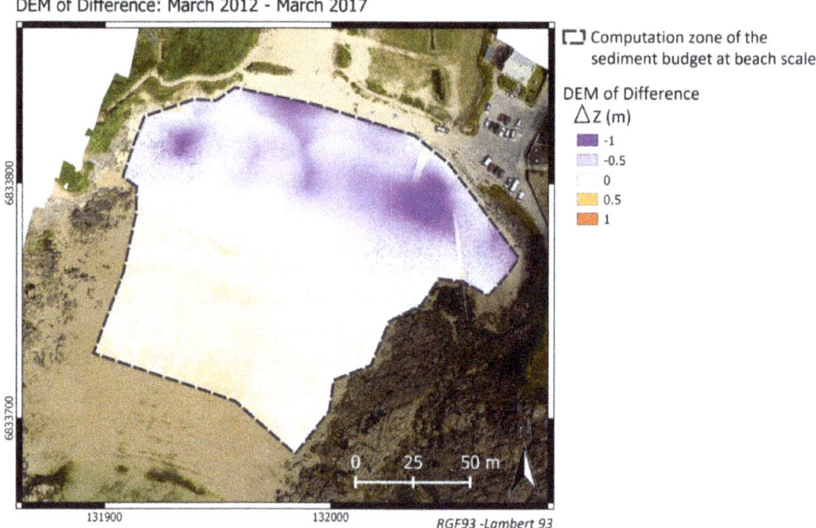

Figure 8. DoDs over the period from March 2012 to March 2017.

Such orders of magnitude in sediment budgets point out the importance of producing high-accuracy monitoring in order to limit the error margins. Furthermore, this example raises several questions about the influence of various parameters, notably, the prior beach state, the proportion of the change in sediment budget that is due to recent variations (during the last few days) or the influence of the dates of survey, and the period considered in the DoD. In Porsmilin Beach, the goal is to produce at least two high-resolution DEMs per year combining UAV and TLS surveys, so as to capture winter and summer periods, as in [30]. For a long-term observatory, despite the flexibility offered by UAV surveys nowadays, it seems difficult, both in terms of the time required for data acquisition and processing and of the required storage capacities, to carry out more than two or three flights per year, especially if more than one site is being monitored. Furthermore, UAV datasets can be combined with other datasets with high spatial resolution and accuracy, such as TLS surveys, which are less restrictive in terms of regulation but more restrictive in terms of spatial coverage.

It appears, therefore, relevant to complement these 3D monitoring techniques to allow capturing spatialized information and high spatial resolution with other monitoring techniques, which would allow capturing smaller temporal-scale morphological variations, such as cross-shore GPS profiles [31] or video imagery systems [32]. Beach profiling indeed enables high repeatability, even in response to storm events. Combining these techniques bridges the observational gap between spatial and temporal resolution and offers a synoptic vision of beach evolution.

5. Conclusions

With drone photographs being collected above Porsmilin Beach since 2006, this long-term monitoring has undergone evolutions in the aircrafts and embedded cameras used as well as an evolution in the processing chain, notably with the emergence of SfM photogrammetry. In parallel, the in situ survey protocol has evolved, drawing from experience but also taking advantage of these technological evolutions. The survey method currently used allows achieving an accuracy of 3 cm (horizontally and vertically) on the DEM and the orthophotograph. UAV photogrammetry appears, therefore, suitable for coastal morphodynamics survey in the context of a long-term observatory. Indeed, it enables rapid acquisition of 3D topographic data at high spatial resolution and low cost. Since the spatial resolution and accuracy are similar to TLS surveys, both monitoring techniques can be equally used, combined, and/or compared. Nevertheless, in a highly dynamic environment, such as a sandy beach, this type of monitoring needs to be combined with monitoring techniques that can capture high-frequency changes, such as GPS profiles or video monitoring.

With the development of drones using RTK GNSS for image geotagging, the survey protocol of Porsmilin Beach will keep evolving in the years to come. This precision positioning will enable a significant reduction in the number of GCPs or even eliminate the need for GCPs, saving time during the in situ survey and the processing step.

Author Contributions: Conceptualization, M.J., C.D., and P.A.; Methodology and Validation, M.J.; Investigation, M.J., J.A., P.G., C.P., V.C., and E.A.; Formal Analysis, M.J. and H.N.; Supervision, N.L.D. and F.F.; Project Administration, L.C.; Writing, M.J. and N.L.D.

Funding: This work is part of the "Service National d'Observation" DYNALIT, via the research infrastructure ILICO. The authors acknowledge financial support provided by ANR project EQUIPEX CRITEX (ANR-11-EQPX-0011). This work was also supported by ISblue project, interdisciplinary graduate school for the blue planet (ANR-17-EURE-0015) and LabexMER project (ANR-10-LABX-19), both funded by the French government (Agence Nationale de la Recherche) under the program "Investissements d'Avenir".

Conflicts of Interest: The authors declare no conflict of interests.

References

1. Baily, B.; Nowell, D. Techniques for monitoring coastal change: A review and case study. *Ocean Coast. Manag.* **1996**, *32*, 85–95. [CrossRef]
2. Klemas, V. Airborne Remote Sensing of Coastal Features and Processes: An Overview. *J. Coast. Res.* **2013**, *287*, 239–255. [CrossRef]
3. Howd, P.A.; Birkemeier, W.A. *Beach and Nearshore Survey Data: 1981–1984 CERC Field Research Facility*; Technical Report CERC-87-9; US Army Engineer Waterways Experiment Station: Vicksburg, MS, USA, 1987.
4. Stive, M.J.F.; Aarninkhof, S.G.J.; Hamm, L.; Hanson, H.; Larson, M.; Wijnberg, K.M.; Nicholls, R.J.; Capobianco, M. Variability of shore and shoreline evolution. *Coast. Eng.* **2002**, *47*, 211–235. [CrossRef]
5. Miller, P.; Mills, J.; Edwards, S.; Bryan, P.; Marsh, S.; Mitchell, H.; Hobbs, P. A robust surface matching technique for coastal geohazard assessment and management. *ISPRS J. Photogramm. Remote Sens.* **2008**, *63*, 529–542. [CrossRef]
6. Long, N.; Millescamps, B.; Guillot, B.; Pouget, F.; Bertin, X. Monitoring the Topography of a Dynamic Tidal Inlet Using UAV Imagery. *Remote Sens.* **2016**, *8*, 387. [CrossRef]
7. Letortu, P.; Jaud, M.; Grandjean, P.; Ammann, J.; Costa, S.; Maquaire, O.; Davidson, R.; Le Dantec, N.; Delacourt, C. Examining high-resolution survey methods for monitoring cliff erosion at an operational scale. *GISci. Remote Sens.* **2017**, 1–20. [CrossRef]
8. Suanez, S.; Garcin, M.; Bulteau, T.; Rouan, M.; Lagadec, L.; David, L. Les observatoires du trait de côte en France métropolitaine et dans les DOM. *EchoGéo* **2012**, *19*. [CrossRef]
9. Nicholls, R.J.; Townend, I.; Bradbury, A.P.; Ramsbottom, D.; Day, S.A. Planning for long-term coastal change: Experiences from England and Wales. *Ocean Eng.* **2013**, *71*, 3–16. [CrossRef]
10. Mahabot, M.-M.; Jaud, M.; Pennober, G.; Le Dantec, N.; Troadec, R.; Suanez, S.; Delacourt, C. Toward a shoreline evolution observatory in tropical environments: The case of back-reef beaches in La Réunion Island. *Comptes Rendus Geosci.* **2017**, *349*, 330–340. [CrossRef]
11. Dehouck, A. Morphodynamique des plages sableuses de la mer d'Iroise (Finistère). Ph.D. Thesis, Université de Bretagne Occidentale, Brest, France, 2006. (In French).
12. Jaud, M. Techniques d'observation et de mesure haute résolution des transferts sédimentaires dans la frange littorale. Ph.D. Thesis, Université de Bretagne Occidentale, Brest, France, 2011. (In French).

13. Jaud, M.; Letortu, P.; Augereau, E.; Le Dantec, N.; Beauverger, M.; Cuq, V.; Prunier, C.; Le Bivic, R.; Delacourt, C. Adequacy of pseudo-direct georeferencing of terrestrial laser scanning data for coastal landscape surveying against indirect georeferencing. *Eur. J. Remote Sens.* **2017**, *50*, 155–165. [CrossRef]

14. Floc'h, F.; Le Dantec, N.; Lemos, C.; Cancouët, R.; Sous, D.; Petitjean, L.; Bouchette, F.; Ardhuin, F.; Suanez, S.; Delacourt, C. Morphological Response of a Macrotidal Embayed Beach, Porsmilin, France. In Proceedings of the 14th International Coastal Symposium, Sydney, Australia, 6–11 March 2016; pp. 373–377.

15. Caulet, C. Les plages sableuses en environnement macro-tidal: De l'influence de la pente sur les processus morphodynamiques. Ph.D. Thesis, Université de Bretagne Occidentale, Brest, France, 2018. (In French).

16. Delacourt, C.; Allemand, P.; Jaud, M.; Grandjean, P.; Deschamps, A.; Ammann, J.; Cuq, V.; Suanez, S. DRELIO: An unmanned helicopter for imaging coastal area. *J. Coast. Res.* **2009**, *56*, 1489–1493.

17. Jaud, M.; Delacourt, C.; Allemand, P.; Grandjean, P.; Ammann, J.; Cancouët, R.; Deschamps, A.; Varrel, E.; Cuq, V.; Suanez, S. DRELIO: Un drone hélicoptère pour le suivi des zones littorales. In Proceedings of the XIèmes Journées Nationales Génie Côtier—Génie Civil (JNGCGC), Les Sables d'Olonne, France, 22–25 June 2010; pp. 485–496.

18. Dehouck, A.; Dupuis, H.; Sénéchal, N. Pocket beach hydrodynamics: The example of four macrotidal beaches, Brittany, France. *Mar. Geol.* **2009**, *266*, 1–17. [CrossRef]

19. Asseline, J.; De Noni, G.; Chaume, R. Note sur la conception et l'utilisation d'un drone lent pour la télédétection rapprochée. *Photo-Interprétation Images Aériennes et spatiales* **1999**, *37*, 3–13. (In French).

20. Jaud, M.; Grassot, F.; Le Dantec, N.; Verney, R.; Delacourt, C.; Ammann, J.; Deloffre, J.; Grandjean, P. Potential of UAVs for Monitoring Mudflats Morphodynamics (Application to the Seine estuary, France). *Int. J. Geo-Inf.* **2016**, *5*, 50. [CrossRef]

21. Lowe, D. Object recognition from local scale-invariant features. *Proc. Int. Conf. Comput. Vis.* **1999**, *2*, 1150–1157.

22. Eltner, A.; Kaiser, A.; Castillo, C.; Rock, G.; Neugirg, F.; Abellán, A. Image-Based Surface Reconstruction in Geomorphometry Merits—Limits and Developments. *Earth Surf. Dyn.* **2016**, *4*, 359–389. [CrossRef]

23. Tonkin, T.N.; Midgley, N.G. Ground-Control Networks for Image Based Surface Reconstruction: An Investigation of Optimum Survey Designs Using UAV Derived Imagery and Structure-from-Motion Photogrammetry. *Remote Sens.* **2016**, *8*, 786. [CrossRef]

24. Jaud, M.; Passot, S.; Allemand, P.; Le Dantec, N.; Grandjean, P.; Delacourt, C. Suggestions to Limit Geometric Distortions in the Reconstruction of Linear Coastal Landforms by SfM Photogrammetry with PhotoScan® and MicMac® for UAV Surveys with Restricted GCPs Pattern. *Drones* **2019**, *3*, 2. [CrossRef]

25. Javernick, L.; Brasington, J.; Caruso, B. Modeling the topography of shallow braided rivers using structure-from-motion photogrammetry. *Geomorphology* **2014**, *213*, 166–182. [CrossRef]

26. Masselink, G.; Castelle, B.; Scott, T.; Dodet, G.; Suanez, S.; Jackson, D.; Floc'h, F. Extreme wave activity during 2013/2014 winter and morphological impacts along the Atlantic coast of Europe. *Geophys. Res. Lett.* **2016**, *43*, 2135–2143. [CrossRef]

27. Scott, T.; Masselink, G.; O'Hare, T.; Saulter, A.; Poate, T.; Russell, P.; Davidson, M.; Conley, D. The extreme 2013/2014 winter storms: Beach recovery along the southwest coast of England. *Mar. Geol.* **2016**, *382*, 224–241. [CrossRef]

28. Garrote, J.; Díaz-Álvarez, A.; Nganhane, H.; Garzón Heydt, G. The Severe 2013–14 Winter Storms in the Historical Evolution of Cantabrian (Northern Spain) Beach-Dune Systems. *Geosciences* **2018**, *8*, 459. [CrossRef]

29. Blaise, E.; Suanez, S.; Stéphan, P.; Fichaut, B.; David, L.; Cuq, V.; Autret, R.; Houron, J.; Rouan, M.; Floc'h, F.; et al. Bilan des tempêtes de l'hiver 2013–2014 sur la dynamique de recul du trait de côte en Bretagne. *Géomorphologie Relief Processus Environ.* **2015**, *21*, 267–292. (In French). [CrossRef]

30. De Sanjosé Blasco, J.; Gómez-Lende, M.; Sánchez-Fernández, M.; Serrano-Cañadas, E. Monitoring Retreat of Coastal Sandy Systems Using Geomatics Techniques: Somo Beach (Cantabrian Coast, Spain, 1875–2017). *Remote Sen.* **2018**, *10*, 1500. [CrossRef]

31. Suanez, S.; Cariolet, J.M.; Cancouët, R.; Ardhuin, F.; Delacourt, C. Dune recovery after storm erosion on a high-energy beach: Vougot Beach, Brittany (France). *Geomorphology* **2012**, *139–140*, 16–33. [CrossRef]
32. Turner, I.L.; Aarninkhof, S.G.J.; Dronkers, T.D.T.; McGrath, J. CZM Applications of Argus Coastal Imaging at the Gold Coast, Australia. *J. Coast. Res.* **2004**, *20*, 739–752. [CrossRef]

 International Journal of
Geo-Information

Article

Combination of Aerial, Satellite, and UAV Photogrammetry for Mapping the Diachronic Coastline Evolution: The Case of Lefkada Island

Konstantinos Nikolakopoulos *, Aggeliki Kyriou, Ioannis Koukouvelas, Vasiliki Zygouri and Dionysios Apostolopoulos

Department of Geology, University of Patras, 265 04 Patras, Greece; a.kyriou@upnet.gr (A.K.); ioannis@upatras.gr (I.K.); zygouri@upatras.gr (V.Z.); apostolopoulos.dionysios@upatras.gr (D.A.)
* Correspondence: knikolakop@upatras.gr; Tel.: +30-2610997592

Received: 23 August 2019; Accepted: 4 October 2019; Published: 30 October 2019

Abstract: Coastline evolution is a proxy of coastal erosion, defined as the wasting of land along the shoreline due to a combination of natural and/or human causes. For countries with a sea border, where a significant proportion of the population lives in coastal areas, shoreline retreat has become a very serious global problem. Remote sensing data and photogrammetry have been used in coastal erosion mapping for many decades. In the current study, multi-date analogue aerial photos, digital aerial photos, and declassified satellite imagery provided by the U.S. Geological Survey (USGS), Pleiades satellite data, and unmanned aerial vehicle images were combined for accurate mapping of the southwestern Lefkada (Ionian Sea, Greece) coastline over the last 73 years. Different photogrammetric techniques were used for the orthorectifation of the remote sensing data, and geographical information systems were used in order to calculate the rates of shoreline change. The results indicated that the southwest shoreline of Lefkada Island is under dynamic equilibrium. This equilibrium is strongly controlled by geological parameters, such as subsidence of the studied shoreline during co-seismic deformation and mass wasting. The maximum accretion rate was calculated at 0.55 m per year, while the respective erosion rate reached −1.53 m per year.

Keywords: photogrammetry; remote sensing; air photos; Pleiades; declassified satellite imagery; UAV; coastline; Lefkada Island

---

1. Introduction

Coastal areas are defined as those where an interface or a transition between land and sea exists. These areas constitute important and sensitive environments with significant complexity, which are, however, difficult to define due to their diversity in form and dynamics as well as to their spatial boundaries [1,2].

Although coastal areas account for 2% of the planet's surface, they contain almost 10% of the Earth's population [3]. Coastal erosion is the wasting of land along the shoreline due to diverse natural or human causes, such as wave action, wave and tidal currents, high winds, earthquakes, landslides, dam construction, drainage, and other human-induced causes of erosion. Coastal urbanization pressure, rising sea levels, and the increasing frequency of storms and cyclones are also maximizing coastal erosion and putting people's lives and properties at risk. Bird (1996) estimated that 70% of sandy shorelines are eroding [4]. However, this research had no robust estimation of shoreline change rates because of the lack of constant and long-term monitoring. Thus, recent studies suggest a lower rate of constant erosion for the world's sandy beaches, in the order of 24% [5].

The usefulness of remote sensing imagery has been practically proven in coastal erosion detection and mapping for many decades. Over the years, remote sensing imagery has been transforming as a result of the evolution of aerial platforms from airplanes and satellites to unmanned aerial vehicles (UAVs). The first attempts to map the coastal zone with aerial photography can be traced back to the 1920s, just after the termination of the First World War, during which the method of photogrammetry was first developed. From the outset, the advantages of aerial photos become obvious: they are easily interpreted, they cover large areas, and they offer the possibility to view and map in 3D and to perform very accurate photogrammetric measurements. In one characteristic example, diachronic aerial images, covering a period of forty years, were photogrammetrically processed for monitoring the shoreline evolution at the southern New Jersey coast [6]. Additionally, aerial photographs were used to examine the role of the presence of *Posidonia oceanica* on shoreline morphology changes in Sardinia, Italy and Corsica, France [7,8].

The next big step was the launch of the first satellites during the early 1970s. The launch of the Landsat series, despite its low spatial resolution, opened new horizons for coastal area mapping due to the existence of multispectral bands, the repeatability of the scenes captured, and the large area coverage. Landsat Thematic Mapper and Enhanced Thematic Mapper images from different years, covering a period of 25 years, were processed in order to map the shoreline changes in Egypt [9]. Landsat imagery covering a 30-year period has been used for the continuous inspection of coastline changes in western Florida [10]. Similar datasets have been interpreted for detecting changes of the shoreline in India [11], China [12], and Namibia [13]. Furthermore, Landsat data, along with Google Earth images and aerial photographs, were processed in order to perform a diachronic survey of the evolution of the coastline in Lower Casamance and southern Gambia, Africa [14]. The development of very high resolution (VHR) satellite sensors (i.e., Ikonos and Quickbird) around 2000 rebooted the use of remote sensing data in coastal monitoring. The new sensors present special characteristics, such as four to eight multispectral bands, programmability in image acquisition globally, high repeatability (between 24 to 72 h), and storage of the data in digital format. An automatic procedure for shoreline extraction from Ikonos images using a mean shift segmentation algorithm was proposed by Reference [15]. More recently, Sekovski et al. [16] processed a WorldView-2 multispectal image to portray the coastline by combining supervised and unsupervised image classification methods. In the last five years, the explosion of the commercial UAV market has opened new avenues in coastal monitoring, as described in Reference [17]. UAVs present some obvious advantages for coastal surveying. These advantages include the best spatial resolution among all the other remote sensing surveys, a lower cost compared to all other ground or remote sensing surveys, potential to transmit the acquired photos to the remote controller, the ability to repeat acquisition in case something goes wrong, and very low security risks as the drones are unmanned.

As mentioned in the previous paragraph, different types of remote sensing data have been used in coastal environment applications. High resolution satellite data and air photos were used to monitor the coastline's changes or to measure volume changes in coastal areas [18–21], sea surface temperatures [22,23], sea ice coverage [24,25], and coastal change processes [26,27]. A review paper summarizing the remote processing methods performed in shoreline delineation has been presented [28]. Recently, the use of UAVs for coastal monitoring was investigated in many studies [29–37]. In one of these studies, UAV digital photogrammetry was applied in order to develop point clouds and 3D models of the coastal cliffs in East Sussex, monthly for a one-year period, in order to map their recession [38]. UAV images and structure from motion (SfM) photogrammetry was also tested at Dongshan Island, China to map coastal changes [39].

In order to map the diachronic evolution of a coastline, there is a need to use multitemporal and multisensor data with diverse spatial resolution. In such a study, air photos and the Landsat and Spot data series covering a period of 45 years were combined in order to map the shoreline evolution in Mexico [40]. Diverse multidisciplinary survey techniques, such as terrestrial laser scanning, terrestrial photogrammetry, and structure from motion photogrammetry using UAV data, have been tried in order to evaluate their potential in cliff erosion monitoring [41]. Aerial photos and very high-resolution satellite data, such as that from Ikonos, Quickbird, Worldview-2, and Pleiades, covering a 14-year period, have been used for estimating/defining coastal erosion in southern Thailand [42]. Topographic maps, Google Earth images, and Landsat Thematic Mapper data were combined to map shoreline changes along the mangrove ecosystem of East Indonesia [43]. A study combining aerial photographs from three different dates (1960, 1973, and 1988) and Quickbird imagery from 2014 focused on the changes in coastline location in the Bay of Jijel (eastern Algeria) [44]. Aerial photographs and a variety of VHR satellite images (Quickbird, Worldview-1, and Worldview-2) were analyzed to map the shoreline changes from 1943 to 2012 in Papua New Guinea [45]. Historical maps, analogue aerial photos, and airborne Light Detection and Ranging (LiDAR) data were compared in order to evaluate the shoreline change for the period 1881 to 2015 [46]. Historical aerial photos and VHR satellite data were processed for shoreline mapping on the Marshal Islands [47,48]. In the context of air photos and satellite images in combination with a geographical information system (GIS), a Digital Shoreline Analysis System (DSAS) and field surveys were utilized for the assessment of coastal erosion and the kinematics of the coastline in order to understand how natural and anthropogenic factors affect the diachronic coastline evolution [49].

In this study, we use a 73-year inventory period motivated by the strong accretion and erosion rates in the southwestern coastal area of Lefkada (Figure 1). In Figure 1, the scale remains at the same position at different dates, indicating the decrease or increase of the beach width. From 1945 to 1972, the erosion rate was quite small; between 1972 and 2008, the erosion increased; and from 2008 to 2016, the phenomenon is reversed and a high accretion rate is observed. Multi-date analogue aerial photos, digital aerial photos, declassified satellite imagery provided by United States Geological Survey (USGS), Pleiades satellite data, and UAV images were combined for the accurate mapping of the western Lefkada coastline during the last 73 years. This is the first time that such a range of data is being analyzed and compared simultaneously. As far as we know, studies on Pleiades triplet data focused mainly on the vertical accuracy of the derived digital surface model and the horizontal accuracy of the produced orthomosaics [50–54]. Only in one case study [42] were Pleiades data combined with other VHR satellite images in order to map shoreline changes. In the current paper, Pleiades triplet data are processed and compared to air photos and UAV data. Another novel characteristic of the current study is the photogrammetric processing of declassified satellite imagery for coastline evolution. Although more than 24 years have passed since the liberalization of these datasets, there are no studies describing their photogrammetric processing. Furthermore, aerial, satellite, UAV photogrammetry, and geographical information systems are handled together for the remote sensing data processing and the shoreline erosion mapping. In addition to the different remote sensing data used in this paper, the geologic environment is equally fascinating as most of the accumulated sediments in the coast come from active landslides. Thus, the specific work tries to correlate the coastline evolution with the earthquakes through time.

The remainder of the current manuscript is divided into five sections: In the next section, the geological and geomorphological status of Lefkada Island is described. Section 3 presents the materials used and the performed methods. In Section 4 the results are stated, and Section 5 discusses and interprets the results. The final section presents the conclusions.

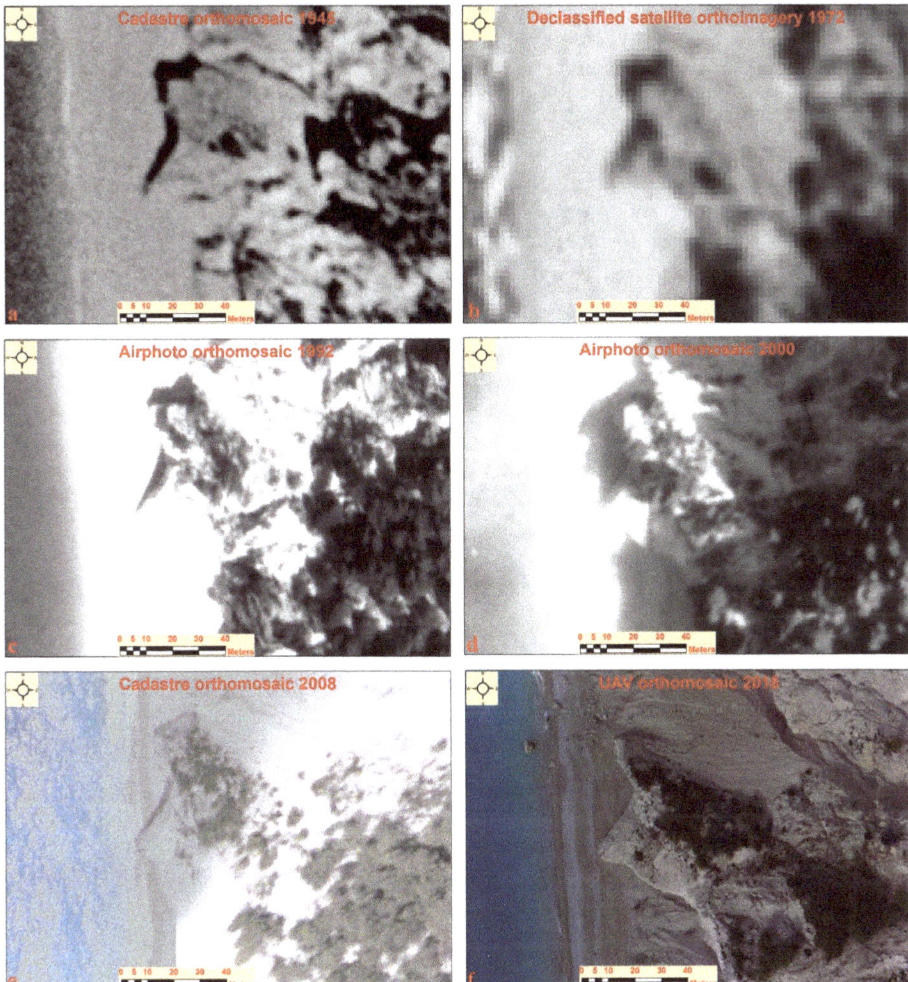

Figure 1. Diachronic images showing a rock outcrop in the central part of the Egremni beach: Note the characteristic erosion and accretion of the shoreline over the years. The scale bar is located at the same point in all the images. (**a**) Air photo mosaic of 1945. (**b**) Declassified satellite imagery of 1972. (**c**) Air photo mosaic of 1992. (**d**) Air photo mosaic of 2000. (**e**) Cadastre mosaic of 2008. (**f**) Unmanned aerial vehicle (UAV) mosaic of 2018.

2. Study Area

The northwestern-most Hellenic arc, where the Ionian Sea islands are located, is regarded as a site of complex continent–continent to continent–ocean convergent plate margins [55]. The collision zone in the north is separated from the continent–ocean convergence margin in the south by the Cephalonia Transform Fault zone (CTFZ). The CTFZ (Figure 2a) is a major dextral strike-slip structure [56], considered as a highly active seismotectonic structure hosting most of the earthquakes in Cephalonia and Lefkada Islands (Figure 2a and Table 1). The fault is separated in a southern segment that strikes SW–NE offshore to the west of Cephalonia Island. The northern segment of the CTFZ strikes SSW–NNE offshore to the west of Lefkada Island. The total length of the CFTZ is about 60 km [57]. Most known large earthquakes in the study area (Table 1) are the cluster of events during the 1953 Ionian earthquakes that seriously damaged the Cephalonia, Zakynthos, Ithaki, and Lefkada Islands [58,59]. Several large earthquakes have also damaged Lefkada Island in the historical period as well as in the instrumental era of seismicity [58]. Many historical earthquakes have also been documented with the largest of them occurring from 1577 up to 1869 AD at magnitudes ranging from 5.9 to 6.5 [60]. The 2003 Mw6.3 [61] and the 2015 Mw6.4 strong earthquakes [62] are regarded as key earthquakes for understanding the seismotectonics of the northern segment of the CTFZ. Most of these earthquakes are large enough and located close enough to Lefkada to trigger significant landslides in the study area [63–65]. Seismicity in the area shows temporal gaps of between 12 and 30 years (Table 1).

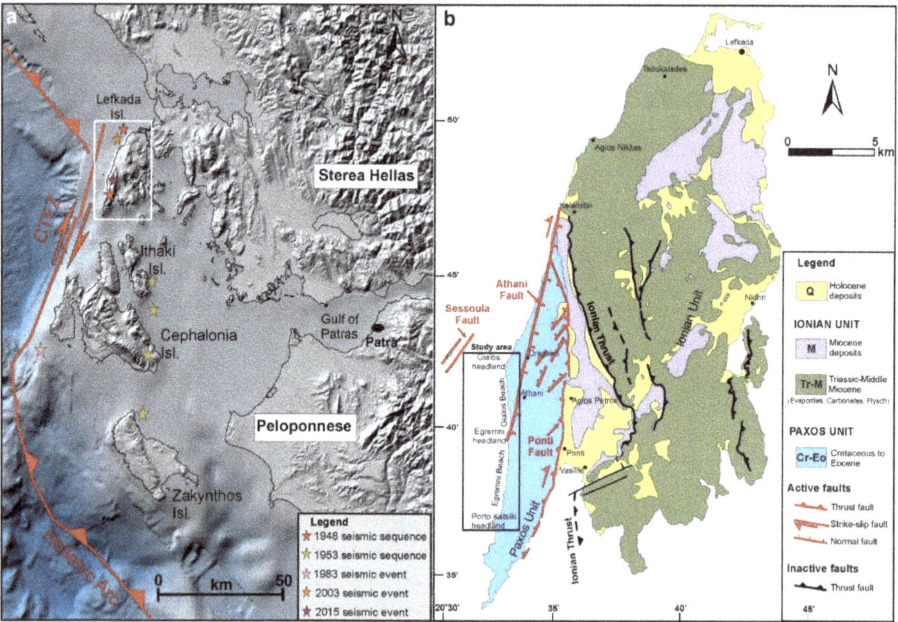

Figure 2. (**a**) The main structural features that affect the seismotectonics of the Ionian Sea: The stars provide the epicenters of the seismic events presented in Table 1. (**b**). A simplified display of the main geological characteristics of Lefkada island: The black box delineates the study area.

In addition to the CTFZ, the so-called Athani Fault is mapped onshore in the study area controlling the western Lefkada coast [66,67]. The Athani Fault is defined for almost 20 km, crossing lengthwise the west coast of Lefkada Island (Figure 2b). The Athani Fault represents a dextral strike slip fault that juxtaposes Paxos unit limestone against Pleistocene to Holocene deposits. In addition to the Athani Fault, other smaller faults, being almost parallel with the CTFZ and Athani Faults with similar kinematics, define steep slopes hosting landslides during earthquake activity [65]. The November 2015 earthquake of Lefkada, 12 years after the 6.3 earthquake of 2003, affected Lefkada's coastal cliffs. The landslides and the coastal area, on which this work focuses, are at obvious structural proximity with two major faults in the area, namely, the Athani Fault and CTFZ. Thus, the western part of Lefkada Island is an ideal study site characterized by high and frequent rates of seismicity (Table 1) and crossed by active faults.

Table 1. The most significant earthquakes affecting Lefkada Island during the studied time period. For more information on the earthquakes see References [62,64,68,69].

Date	Longitude	Latitude	Depth	Mw
22 April 1948	38.68	20.60	Unknown	6.5
30 June 1948	38.80	20.60	Unknown	6.4
12 August 1953	38.30	20.80	Unknown	7.0
30 Years of No Seismicity				
17 January 1983	37.97	20.25	9 km	6.5
20 Years of No Seismicity				
14 August 2003	38.79	20.56	12 km	6.3
12 Years of No Seismicity				
17 November 2015	38.67	20.60	11 km	6.5

3. Materials and Methods

3.1. Materials

In the current study, different datasets of airborne, satellite, and UAV platforms were combined and processed using diverse photogrammetric techniques. Acquisition dates, number of data, spatial resolution, and other characteristics of the datasets are presented in Table 2.

Table 2. The datasets used in this study, their source, and their spatial resolution.

Date	Data Type	Source	Reference System	Number of Photos	Spatial Resolution
1945	Air photo mosaic	National Greek Cadastre and Mapping Agency	Greek Grid	Orthomosaic	1
27/05/1972	Declassified satellite imagery	United States Geological Survey	No reference system	1	2 m
05/06/1974	Declassified satellite imagery	United States Geological Survey	No reference system	1	4 m
05/10/1980	Declassified satellite imagery	United States Geological Survey	No reference system	1	4 m
11/09/1992	Analogue air photos	Hellenic Military Geographical Service	No reference system	8	1 m
30/05/2000	Analogue air photos	Hellenic Military Geographical Service	No reference system	8	1 m
2008	Digital colored air photos	National Greek Cadastre and Mapping Agency	Greek Grid	Orthomosaic	0.5 m
2016	Pleiades Imagery	Enceladus Greek Supersite	World Geodetic System 84	2 triplets	0.5 m
3/11/2018	Unmanned Aerial Vehicle Imagery	University of Patras	World Geodetic System '84	462	0.1 m

The first historical dataset is an orthophoto mosaic of 1945 created for the needs of the Greek Cadastre. The specific digital orthomosaic was created with photogrammetric methods from analogue aerial photographs acquired in 1945. Having a pixel size of 1 m, it covers the whole Greek territory.

The specific dataset was developed by the National Greek Cadastre and Mapping Agency. We did not perform any further process on it.

The second dataset includes 3 sets of declassified satellite imagery obtained in the late 1960s–1970s by American military missions and declassified for the first time after 1995. The specific declassified photographs were acquired by U.S. military satellites and give globally important information of the planet surface [70]. This imagery was, for many years, classified as top secret, and it was delivered for free use in 1995. The specific archive contains more than 990,000 photographs. The photos were acquired between 1959 and 1980. The images present a variety of scales and quality. Very often, the scenes are cloudy. The pixel size of the photos varies and ranges between 6 and 30 feet (2–9 m). The film is scanned at 7 microns, and the final total size of an image can surpass 1.3 GB [71]. The first image of the specific dataset was collected during the CORONA KH-4B mission in May 1972 (Table 2). The specific satellite collected the photographs with a telescopic camera, and the film returned to the earth via recovery capsules. While the first Corona missions carried a single panoramic camera or a single frame camera, the next satellites (KH-4, KH-4A, and KH-4B) had double panoramic cameras on board, each looking at 15 degrees forward or backward from the satellite orbit. The pixel size of the image was around 6 feet, but according to previous studies, the KH-4B missions provided the best spatial resolution (1.83 m at nadir [72,73]. The second and the third images of the specific dataset were collected during the CORONA KH-9 program missions. The KH-9 program was active for 7 years (1973 to 1980). The specific mission collected images according to a predefined grid in order to eliminate the image distortion. The images were acquired with overlap for stereoscopic analysis. The KH-9 system produced 9 × 18 inch photos with a pixel size of 20 feet. During the 7-year mission, KH-0 collected almost 29,000 images in 12 accomplished space journeys. The second image of the declassified satellite imagery dataset was acquired in June 1974 during the mission 1208-5. The third image was acquired in October 1980 during the mission 1216-5. The spatial resolution of both images is better than 20 feet, and they were both scanned at 7 microns.

The third dataset comprises 8 analogue aerial photographs from 1992, at 1/15,000 scale accessed through the Hellenic Military Geographical Service (HMGS). With 60% along the track overlap, these photos developed 7 stereopairs and were selected as they combine the best spatial resolution and are cloud free.

The fourth dataset comprises eight analogue aerial photographs from 2000, at the same scale (1/15,000), and from the same source (HMGS).

During the period 2007–2009, digital air photos covering the whole of Greece were acquired by the National Greek Cadastre and Mapping Agency. After photogrammetric processing, an orthomosaic with a pixel size of 50 cm and a digital surface model with a spatial resolution of 5 m were produced. Both covered the whole country, and they are used as basemaps in many studies as they present the highest horizontal and vertical accuracies. The orthophoto and the Digital Surface Model of National Greek Cadastre and Mapping Agency covering Lefkada Island belong to the fifth dataset used in this study.

The sixth dataset consists of 2 scenes of Pleiades satellite imagery. The Pleiades system comprises two satellites in the same orbit but with 180° offset. The first satellite was launched in 2011, while the second satellite was launched 1 year later. Each satellite simultaneously collects 1 panchromatic and 4 multispectral bands. The panchromatic band has a pixel of 0.7 m, while the 4 multispectral bands have a pixel size of 2.8 m [74]. Lefkada Island is covered by two scenes (triplets). The first scene was acquired on 26 September 2016, and the second was acquired on 13 October 2016. Each scene (triplet) was composed of 3 images (tri-stereo), and both scenes were totally free of clouds.

A total of 462 UAV photographs comprise the seventh and final dataset. These were acquired by the University of Patras team during field work in November 2018. A photogrammetric flight was performed in the Egremni beach area. The flight's altitude was 60 m above ground level (agl). The along-the-track overlap of the photos was 90%, while the respective across-the-track overlap was 75%. Table 3 presents all the flight details.

Table 3. Characteristics of the UAV flight campaign and pixel size of the produced orthophoto and Digital Surface Model.

Flight Campaign Altitude (m)	Along the Track Overlap %	Across the Track Overlap %	Number of Photos	Orthophoto Pixel Size (m)	DSM Pixel Size (m)	Image Quality (pixel)
60	90	75	462	4.2 cm	8.4 cm	3000 × 4000

3.2. Methods

The 7 datasets noted in the previous paragraph were processed using different photogrammetric and computer vision techniques discussed below. We used two software packages, ERDAS IMAGINE Leica photogrammetry Suite (LPS) and Agisoft Photoscan Professional, for the processing of the various remote sensing data. The Greek Cadastre orthophoto and Digital Surface Model (DSM) of 2008 were used as a basemap and reference for all the other datasets. According to the Greek Cadastre specifications, the orthomosaics have a horizontal (planimetric) accuracy of 0.71 m while the vertical accuracy of the respective DSM is better than 2 m.

3.2.1. Analogue Air Photo Data Processing

In LPS (2014 release), a block was created and the analogue air photos were imported into it (Figure 3). The internal triangulation of the air photos was calculated using the fiducial points on each photo. Furthermore, there is a need for external Ground Control Points (GCPs) to orthorectify the imagery (Figure 3). There are many possible error sources in an air photo block, such as lens distortion, film distortion, and atmospheric refraction [75]. Each of these may possibly decrease the accuracy of aerial triangulation results and the accuracy of the final orthomosaic. The major parameters which increase or decrease the accuracy of the triangulation and the accuracy of the final orthomosaic are the following: the position of the air photos (block geometry), the allocation, the quantity and the accuracy of the points that are used (control or tie points), and the existence of any other random error. The whole procedure is described in detail in a previous study [75] and cannot be repeated in the current paper. As the block geometry and the existence of random errors are predefined, the main duty of the user is to select with high precision the GCPs and the tie points.

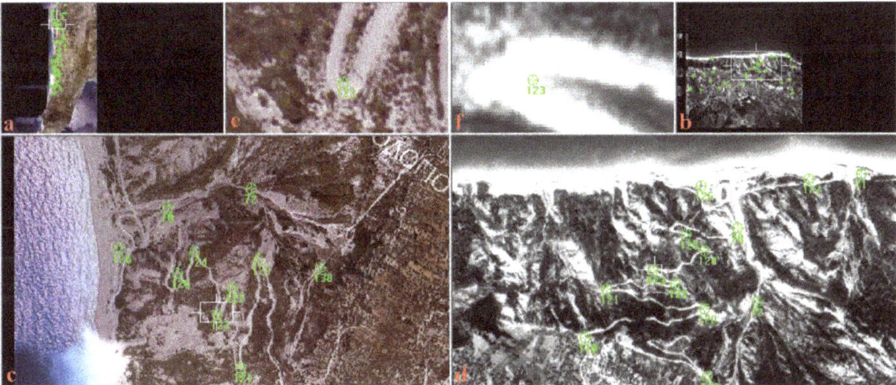

Figure 3. Selection of ground control points during the orthorectification process of the air photos of 2000: (**a**) Cadastre orthomosaic displaying the study area. (**b**) One of the air photos of 2000 displaying the study area. (**c**) Enlargement of Figure 3a. (**d**). Enlargement of Figure 3b. (**e**). Focus on one ground control point on the Cadastre orthomosaic. (**f**) Focus on the respective ground control point on air photo of 2000.

In the first LPS block of analogue air photos of 2000, we used 71 GCPs, while in the second LPS block of 1992 air photos, we used 54 GCPs. The coordinates of GCPs were obtained from the Greek Cadastre orthomosaic, while the respective altitude was obtained from the respective DSM. The GCPs had a very good distribution around the broader area with an emphasis given on the area across western Lefkada beach (Figure 3). Then, LPS was used to calculate the root mean square error (RMSE) values for the entire block, as seen in Table 4. The RMSEs calculated for the two blocks are assumed acceptable in order to proceed to the orthorectification of the photos and the creation of the final orthomosaics, one for the May 1992 air photos and one for the September 2000 air photos.

Table 4. The number of ground control points used in each Leica Photogrammetry Suite block, the respective number of tie points, and the aerial triangulation root mean square errors.

LPS Block	No of Images	No of Ground Control Points	No of Tie Points	RMSE (m)
Declassified 1972	1	58	0	0.0044
Declassified 1974	1	45	0	0.0727
Declassified 1980	1	38	0	0.1668
Air photo 1992	8	54	83	1.9753
Air photo 2000	8	71	92	1.8602
Pleiades 2016	6	180	224	0.1263

3.2.2. Declassified Satellite Imagery Processing

The 3 declassified satellite scenes were processed in 3 different blocks in LPS, one for each date. In a previous study [76], LPS was used for the analyses of CORONA KH-4 imagery. That study demonstrated the huge contribution of these data process for archaeological studies in the Near East. Stereo models, digital elevation models (DEMs), and orthomosaics have been successfully produced utilizing a simplified frame-model in LPS for orthorectification of the CORONA KH-4 images. The GCPs used in that study [76] were selected from freely available datasets. LPS was also used for the orthorectification of CORONA KH-4 data in another study [77]. Ikonos Geo product imagery and Shuttle Radar Topography Mission DEMs were used for the orthorectification of the declassified image. CORONA imagery has been processed in ERDAS IMAGINE OrthoBASE Pro in order to produce a DEM [78]. Sohn et al. [73], among others, used the same software to assess the quality of CORONA imagery derived, and Reference [79] proposed a methodology for CORONA image processing based on ERDAS IMAGINE toolboxes. CORONA KH-9 imagery was processed with satisfactory results in the past [71,80].

In the current study, the 3 declassified scenes were orthorectified using LPS (2014 release). Around 50 points were selected from the Greek Cadastre orthophoto to serve as the GCPs for each declassified scene. Particularly for the 1972 imagery, 58 ground control points allocated all over the scene were used (Figure 4). The declassified imagery of 1972 was orthorectified with a pixel size of 2 m, while the respective images of 1974 and 1980 were orthorectified with a pixel size of 4 m.

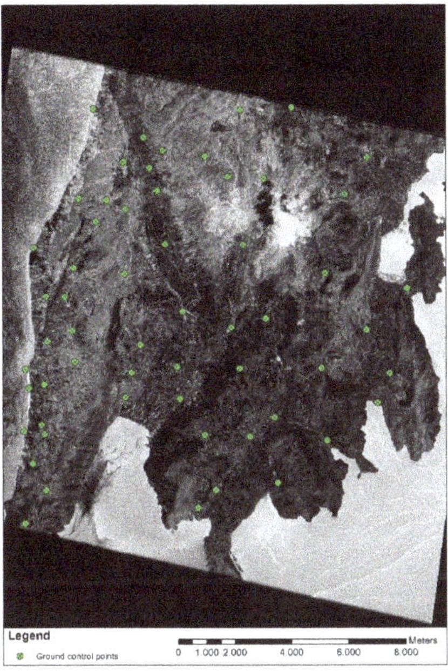

Figure 4. Allocation of the ground control points on the declassified satellite imagery.

3.2.3. Pleiades Data Processing

Pleiades tri-stereo images were processed using also LPS. The exact procedure for Pleiades triplets processing is described in Reference [54]. Initially, all 6 panchromatic images that group the two triplets were inspected for radiometric unconformities during the image acquisition as proposed in Reference [54]. Commonly, radiometric anomalies show up only in one of the two or three images (stereo pair or triplet). They are caused by the existence of surfaces of highly reflective objects (e.g., metal roofs and water surfaces). A different acquisition angle may produce sun glint in one of the images and not in others [50]. None of these effects was noted in the present set of Pleiades images. The coordinates of 180 GCPs were obtained from the orthomosaic provided by the National Cadastre and Mapping Agency S.A., while the elevation was retrieved by the respective DSM.

3.2.4. UAV Data Processing

UAV images were imported into the Agisoft PhotoScan software. The software combined computer vision techniques and structure from motion (SfM) photogrammetry to achieve direct georeferencing or bundle adjustment with ground control points (GCPs) [75,81,82] or simple similarity transformation over the whole block without GCPs. Alike to classic photogrammetry from air photos or satellite stereopairs, SfM photogrammetry uses overlapping images taken from different points of view. The major variation between the two approaches is that SfM defines the internal camera characteristics, the position of the camera, and the orientation of each image in an automatic way without the need for any prior knowledge or grid determination [83]. All the necessary calculations for the internal orientation are estimated automatically by a repeat procedure called "bundle adjustment". Bundle adjustment reflects the automatic localization of common characteristics in a group of overlapping photos [75]. The geometry of the whole scene is built as more overlapping images are processed and more mutual objects are detected and related. The requirement for many overlapping photos in order to cover an entire area of interest generated the procedure name of structure from motion photogrammetry or photogrammetry produced from a moving camera [82].

During the flight campaigns, artificial targets were spread out in the broader area and measured with a differential Global Navigation Satellite System (GNSS) receiver (Leica GS08). These artificial targets were easily detectable in the UAV images and used as ground control points. The use of artificial targets as ground control points is described in detail in a previous study [17]. The allocation of these targets is presented in Figure 5. As described in more detail in another case [75], the GNSS sensor was receiving corrections—through the GSM network—from the Greek Hellenic Positioning System (HEPOS). The receiver collects signals from GPS (L1, L2, and L2C frequencies), GLONASS (L1 and L2 frequencies), and Satellite-Based Augmentation Systems like Wide Area Augmentation System (WAAS), European Geostationary Navigation Overlay Service (EGNOS), etc. All the GCPs' measurements have a horizontal accuracy better than 1.3 cm and a vertical accuracy better than 2 cm.

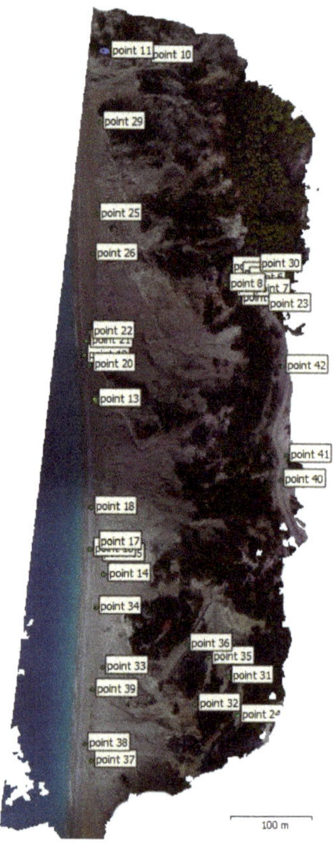

Figure 5. Allocation of the ground control points measured with differential Global Navigation Satellite System (GNSS) receiver on the UAV orthomosaic.

3.3. Historical Shoreline Analysis

To quantify shoreline changes, we separated the 73-year period from 1945 to 2018 into 3 subperiods. These were 1945–1972, 1972–2000, and 2000–2016. This time separation enabled the detailed analysis of the erosion and accretion in rates and in absolute lengths highlighted along the shoreline trace. For the shoreline analysis, two different approaches were used. Firstly, the shoreline was digitized from all the datasets. Then, two different flowcharts—one fully automatic and one semiautomatic—were followed for the shoreline evolution mapping, as described in the next paragraphs. The decision to use two different flowcharts was based on a previous study [84] which proved that different techniques can provide very correlated results for smooth shorelines but less correlated results for irregular shorelines.

3.3.1. Digital Shoreline Analysis System (DSAS)

A software package called the Digital Shoreline Analysis System (DSAS) was used in order to measure the shoreline rates-of-change by comparing vectorized shorelines from diverse dates [85]. DSAS is an add-on tool of ArcGIS software, developed by the USGS in conjunction with the TPMC Environmental Services. The DSAS software creates transect lines perpendicular to the coastline with reference to a specific baseline set by the user. The user also determines the spacing of the transect lines across the shoreline [40]. The shift of the shoreline seaward or landward, with reference to the baseline, is characterized as accumulation or erosion at each transect, respectively, and the statistical values are considered as positive for accumulation and negative for erosion. The software calculates diverse statistical values in order to measure the shoreline position change [85]. The main values are the end point rate (EPR), which measures the rate of the coastline change between two successive shorelines, and the net shoreline movement (NSM), which calculates the total distance between the successive shorelines. The specific method for coastline change detection has been used in the past with different types of remote sensing data. For example, DSAS was used with multi-date satellite images from the Indian sensors (IRS P6 and LISS-III) in order to extract the shoreline change on the western coasts of India [86]. The same software was used with diverse Landsat Thematic Mapper and Enhanced Thematic Mapper images during a 25-year period to derive the shoreline changes in Egypt [7]. In another example, diachronic aerial photographs and Quick-bird satellite data were used in combination with DSAS [44] in order to detect and measure the shoreline movements at the Bay of Jijel (eastern Algeria). Multitemporal and satellite dataset have been interpreted with DSAS for the change detection in 112 km of shoreline in India [11]. Aerial photographs and a variety of VHR satellite images (Quickbird, Worldview-1, and Worldview-2) were analyzed to map the shoreline changes from 1943 to 2012 in Papua New Guinea [45] with DSAS. Multi-date Landsat images were also processed with the same software [12]. Historical aerial photos and VHR satellite data were processed for the shoreline mapping in the Marshal Islands using DSAS [47,48].

In the current study, DSAS utilizes the digitized shorelines in reference to a baseline projected to the same reference system (Greek Grid). The shorelines cover the following years: 1945, 1972, 1974, 1980, 1992, 2000, 2008, 2016, and 2018. For each shoreline vector, the software requires that the date is predefined in the format year/month/day. The software creates transects which intersect the multi-date shorelines at specific points, which are used for the calculations of the EPRs. The EPR calculation is presented in Figure 6, while the general flowchart of DSAS is presented in Figure 7.

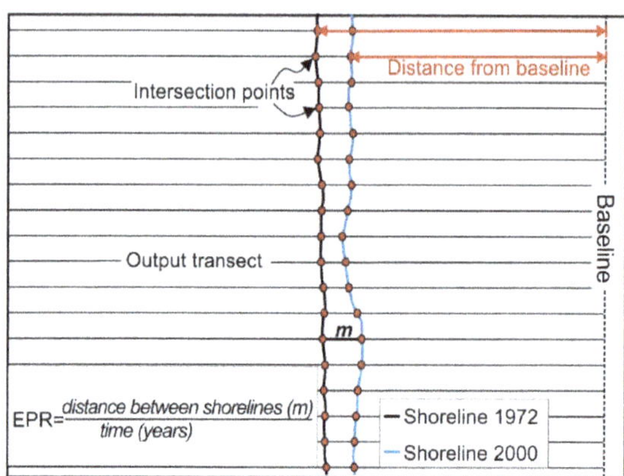

Figure 6. An example of the end point rate (ERP) calculation using 1972 and 2000 shorelines at the study area.

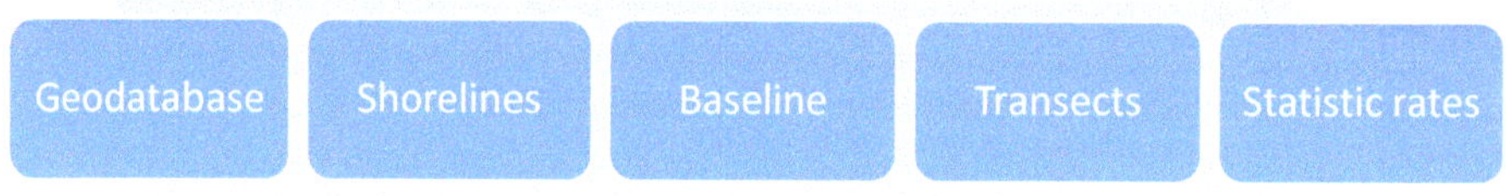

Figure 7. Flowchart of the algorithm of Digital Shoreline Analysis System (DSAS) v4.3.

3.3.2. Semiautomatic Processing in GIS Environment

For the shorelines of Egremni and Gialos in SW Lefkada Island, we also applied semiautomatic processing in a GIS environment for extracting shoreline alterations through time. The methodological procedure was based on the creation of 102 transect lines spaced every 100 m. Their contact point with the studied shorelines identifies the coastal point viewed in a distance diagram. The baseline, which was the road from Porto Katsiki to Athani and from Athani to Komilio that can be identified in images more recent than 1972, was projected to images of all different ages. In Figure 8, the digitized road from the 1972 orthophoto is overlaid in the orthomosaics of 2000 and 2016. The digitized roads confirm the accuracy of the derived orthomosaics of different ages because their spatial adjustment in all the orthomosaics can be identified easily and they also act as testimonies of the used methodology (Figure 8). Thus, the semiautomatic procedure used in this paper represents an evaluation tool for the DSAS methodology and attaches great importance in the study to the shorelines of Egremni and Gialos of Lefkada Island in terms of accretion and erosion.

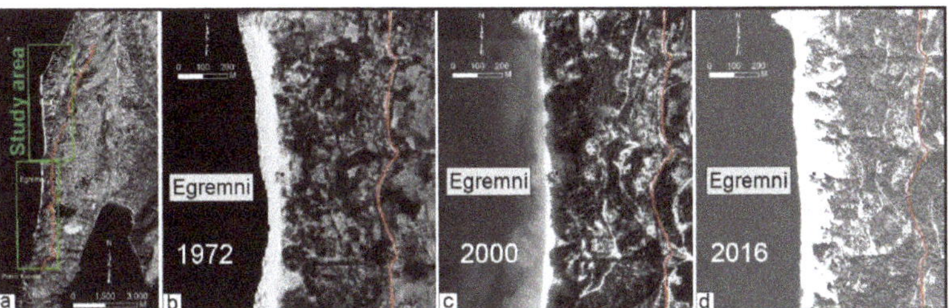

Figure 8. (**a**) The study area where the DSAS and semiautomatic processing in geographical information system (GIS) were performed: Orthomosaics from (**b**) 1972, (**c**) 2000, and (**d**) 2016. In red color, a digitized road from the 1972 orthophoto is overlaid in all mosaics.

4. Results

The images of 1945, 1972, 1974, 1980, 1992, 2000, 2008, and 2016 and a UAV flight campaign during 2018 were used to analyze the rate of change at two significant and popular tourist beaches, Egremni and Gialos, in which both include similar sandy and rocky parts (Figure 2b). These shorelines are regarded as significant for the geomorphological evolution of the west part of the island because a series of factors seem to have a common impact on their evolution. These factors are the wave action; the seismicity; and, for the years after 2000, the human factor. The 73 years of inventory of both beaches revealed remarkable changes over time across the shoreline in the previously referred periods. In these periods, we mapped high erosion and/or accretion rates and absolute length changes. We also defined shoreline trends from erosion to accretion and vice versa (Figure 9). Analytically, our results over discrete periods are summarized in Figure 9. In total, the shoreline mapping results are presented in three figures (Figures 9–11) and in one table where the maximum erosion and accretion rates are highlighted in each time period (Table 5). In Figures 10 and 11, maximum erosion (up to 149 m) and maximum accretion (up to 44 m) can be observed. Over the entire coast, from the Porto Katsiki headland to the northernmost end of the studied shoreline (Gialos headland), erosion affects 6300 m and accretion affects 3700 m, indicating that erosion has prevailed over accretion during the last 73 years.

For the period of 1945–1972, erosion was dominant in six areas where the erosion annual rate surpasses 0.50 m and three large areas where accretion surpasses 0.50 m per year (Figure 9a). Overall, 4800 m of the coast is under erosion while accretion extends to 5200 m, suggesting that the area of erosion is almost equal to the area of accretion. In the south (Egremni beach), accretion in the period 1945–1972 shows a maximum of 51 m while maximum erosion is 54 m (Figure 10a). In the Egremni beach, the accretion is active in 2800 m, or 60% of the total length of the beach, with erosion prevailing in 1900 m, or 40% of the beach length. For the same period in the Gialos beach, accretion is active in 2400 m, or 45% over the beach length, and erosion in 2900 m, or 55% of the beach length (Figure 11a). Maximum accretion in Gialos beach is as much as 22 m and maximum erosion is 30 m.

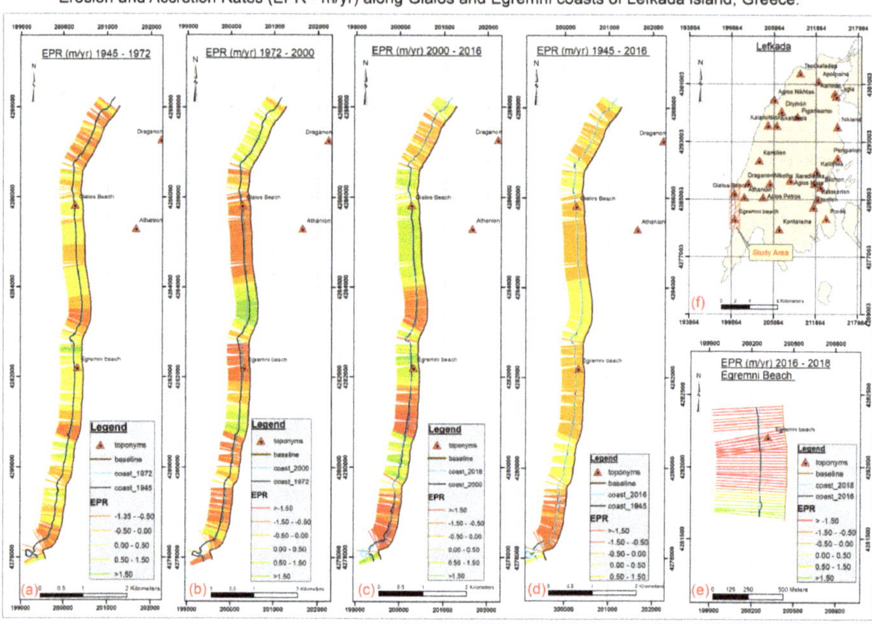

Figure 9. The processing results of DSAS software for the study area: Red colors represent erosion rate per year, while green colors represent accretion rate per year. (**a**) Six areas where the erosion annual rate overpasses 0.50 m and three large areas where accretion overpasses 0.50 m per year for the period of 1945–1972. (**b**) Erosion into areas of accretion and vice versa for the period of 1972–2000. (**c**) One more change from the erosion to accretion trend in comparison with the previous period (1972–2000). (**d**) Areas where erosion and accretion take place over the period of 1945–2016. (**e**) Highest erosion overpasses 1.50 m per year observed in the southernmost end of the Egremni beach for the period of 2016–2018. (**f**) Study area in Lefkada Island.

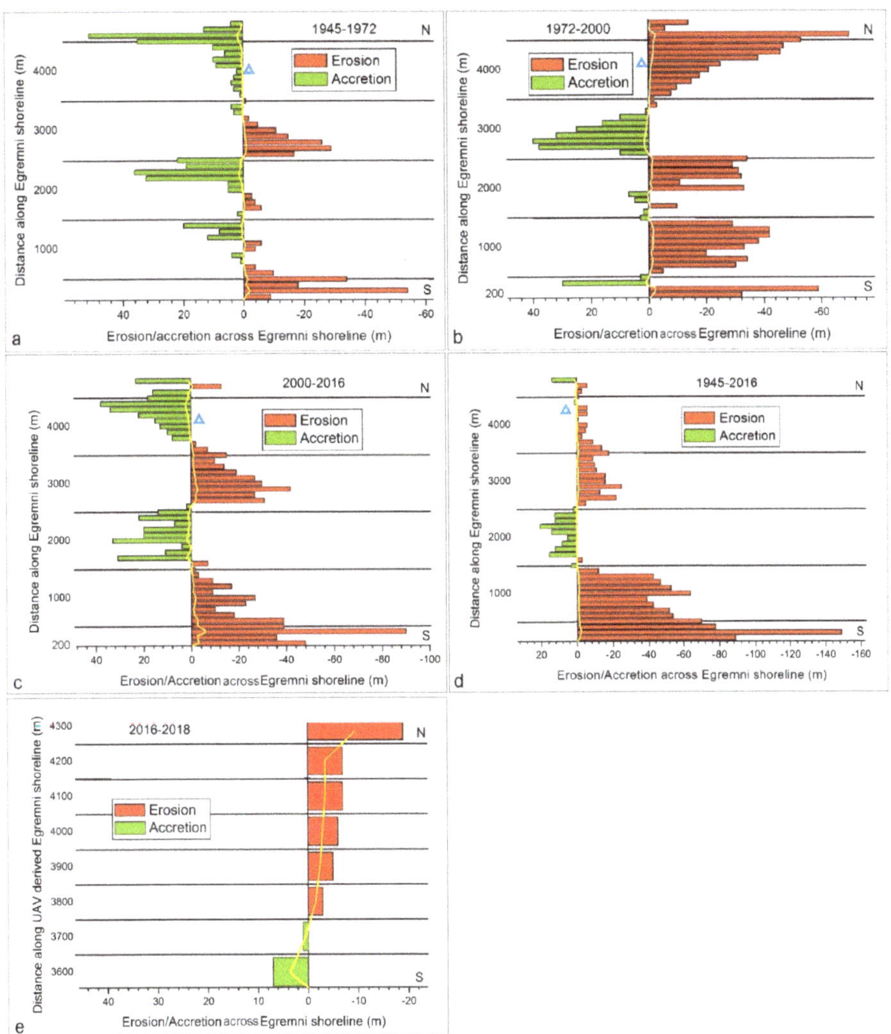

Figure 10. The processing results of the semiautomatic method for Egremni beach: Red colored bars represent total erosion in each transect for the specific time period, while green colored bars represent total accretion for the same time period. The blue triangle states the Egremni rock outcrop shown in Figure 1. The yellow line represents the erosion/accretion rate along the Egremni beach for each time period. N and S are abbreviations for north and south compass points. Each diagram represents a specific time period: (**a**) 1945–1972 period, (**b**) 1972–2000 period, (**c**) 2000–2016 period, (**d**) 1945–2016 period, and (**e**) 2016–2018 period.

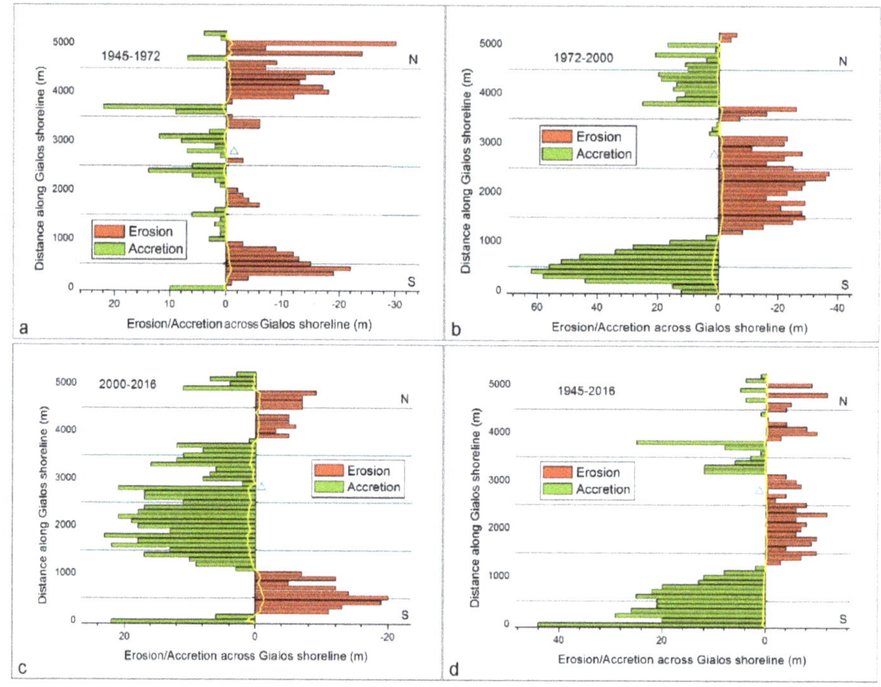

Figure 11. The processing results of the semiautomatic method for Gialos beach: Red colored bars represent total erosion in each transect for the specific time period, while green colored bars represent total accretion for the same time period. The blue triangle states the Gialos reference in Figure 9. The yellow line represents the erosion/accretion rate along the Gialos beach for each time period. N and S are abbreviations for north and south compass points. Each diagram represents a specific time period: (**a**) 1945–1972 period, (**b**) 1972–2000 period, (**c**) 2000–2016 period, and (**d**) 1945–2016 period.

Table 5. Maximum and minimum rates of EPR along the study area.

	Maximum/Minimum Rates of EPR (DSAS Method)		Maximum/Minimum Rates of Accretion/Erosion (Semiautomatic Processing Method)			
			Egremni Beach		Gialos Beach	
Period	Maximum Rate (m/yr)	Minimum Rate (m/yr)	Maximum Rate (m/yr)	Minimum Rate (m/yr)	Maximum Rate (m/yr)	Minimum Rate (m/yr)
1945–1972	1.92	−1.35	1.89	−1.26	0.81	−1.11
1972–2000	2.45	−2.53	1.07	−2.10	2.21	−1.32
2000–2016	2.24	−5.73	2.38	−5.62	1.44	−1.25
1945–2016	0.55	−1.53	0.21	−2.10	0.62	−0.17

During the period 1972–2000, we defined the change from erosion into accretion in comparison to the 1945–1972 period. The new pattern depicts erosion into areas of accretion and vice versa (Figure 9b). At the southern end of the shoreline, erosion prevailed in both periods but now has an impact on a wider area (Figure 10b). At Egremni beach during this period, erosion is remarkably high, up to 70 m in the north and up to 38 m in the south (Figure 10b). Overall, the accretion is active in 1400 m, or 30% of the total beach length, while erosion is prevailing in 3300 m, or 70% of the beach length (Figure 10b). In the Gialos beach, accretion is active in 2800 m, or 53% of the beach length, and erosion in 2500 m, or 47% of the beach length (Figure 10b). Maximum accretion in Gialos beach is as much as 62 m, and maximum erosion reaches 37 m (Figure 11b).

Over the 2000–2016 period, there was again a change from an erosion to an accretion trend in comparison with the previous period (1972–2000). The new pattern once more shows erosion moves into areas of accretion and vice versa (Figure 9c). The southernmost end of the shoreline remains constantly under erosion in this period (Figure 9c). At the Egremni beach during this period, erosion is remarkably high, up to 42 m in the north and up to 90 m in the south (Figure 10c). Maximum accretion is 38 m at the north end of the Egremni beach (Figure 10c). Overall, the accretion is active in 2000 m, or 42% of the total length of the beach, while erosion is prevailing in 2700 m, or 58% of the beach length (Figure 10c). At the Gialos beach, the accretion affected an area 3400 m long, or 64% of the beach length, and erosion affected 1900 m, or 36% of the beach length, in the period 2000–2016 (Figure 11c). Maximum accretion in Gialos beach is as much as 22 m, and maximum erosion is 20 m. In summary, over the 73 years, both beaches were under erosion, with the highest erosion observed at the southernmost end of the Egremni beach near the Porto Katsiki headland (149 m) (Figure 10d), and the highest accretion was observed at the southern end of Gialos beach near Egremni headland (44 m) (Figure 11d).

During the 73 years, there are periods related to intense seismicity and periods with low or absence of seismicity (Table 1). Significant seismicity is recorded at the beginning of the first period, during 1948 and 1953. In the second period, the biggest earthquake occurred in the middle of the 1972–2000 period. The most recent period of seismicity was recorded in 2003 and 2015. Based on these data, our mapping of shorelines corresponds to significant, low, and very high seismicity. Based on our unpublished mapping data relating to the 2003 and 2015 earthquakes, significant landslides and mass wasting across the shoreline were triggered due to these two earthquakes.

5. Discussion

The phenomenon of shoreline changes is quite complex, affected by oceanographic and meteorological conditions and geological parameters. In general, the study area is influenced by west to northwest prevailing winds that rarely blow above a maximum speed of 6 Bf [87]. The tide wave action is characterized by low amplitude as in the rest of Mediterranean Sea, and the average offshore wave height is almost 0.79 m [88]. However, in this study, as oceanographic and meteorological data and studies are missing for Lefkada Island, we have to focus exclusively on geological parameters such as seismicity, active deformation, and mass wasting across the west cliff of Lefkada in order to explain the shoreline changes. Those changes were mapped using multi-date and multisensor remote sensing data combining, for the first time, aerial photos, declassified satellite imagery, Pleiades data, and UAV data. The photogrammetric processing of these data and the very low RMSE (Table 4) resulted in the excellent georeferencing of the diachronic data (Figure 8). The extensive field work provided the necessary number of ground control points for the accurate georeferencing of the UAV data in order to combine them with the other remote sensing data. This is fully in accordance with previous studies [89,90] that noted that the registration of UAV imagery with any other remote sensing coarser product depends on the use of ground control points that succeed in highly accurate georeferencing. However, the requirements of this procedure are by far more demanding in time and ground support [89].

The study area includes two sweeping beaches, located at southwestern Lefkada Island, separated by a headland developed between the Gialos and the Egremni beaches (Egremni headland). To the south, the study area ends at the Porto Katsiki headland and, to the north, at Gialos headland. The beach to the north of the Egremni headland is characterized by maximum accretion of 44 m (Figure 11d), while north of the Porto Katsiki headland, erosion dominates, reaching up to 149 m in the time period from 1945 to 2016 (Figure 10d). Since erosion or accretion is constantly sustained through the 73-year period in these two locations, we consider that the beach topography and oceanographic conditions are the main mechanisms controlling erosion or accretion at these specific sites. In contrast to these sites, all other shoreline lengths show remarkable changes from erosion to accretion and vice versa. Trying to highlight these changes in a better way, we considered seismicity as a major factor controlling sediment equilibrium and active deformation through earthquake induced landslides along the Egremni–Gialos beaches.

Indeed, the study of different time imageries confirms that the landslides in the overhanging cliff are directly related to earthquake occurrence and affect the forest development or retreat (Figure 12). Figure 12 presents the diachronic evolution of one of the landslides in the overhanging cliff in Egremni beach. It is obvious that from 1945 to 2018 that the extent of the landslide was almost doubled, offering waste material to the Egremni beach. The usual material that covers Lefkada beaches is mostly coarse sand with a small percentage of pebbles [87]. However, both Egremni and Gialos Beaches present a significant percentage of medium to large pebbles and granules (Figure 12) attributed to mass wasting phenomena evolving in both beaches. Thus, our multidate analysis indicates that, primarily, the seismic events of 1948, 2003, and 2015 [69] intercalated in two out of three time periods that represented crucial parameters for cliff erosion and sediment accumulation along the studied shoreline. However, Reference [69] indicated that the analyzed shoreline during the most recent seismicity has subsided a few centimeters and Ganas et al. 2016 [91] suggested extensive coastal road failures related to the outbreak of numerous earthquake-induced landslides. Thus, the changes of the two periods 1945–1972 and 2000–2016 are regarded as the result of the strong earthquakes of 1948, 1953, 2003, and 2015 (Figure 2a and Table 1). Their immediate impact is the prevalence of shoreline accretion, especially in the sandy part of both beaches (Figures 10a,c and 11a,c). In contrast, during the 30-year period of seismic quiescence (from 1953–1983) and especially in the 1972–2000 period, only one earthquake affected the area (Figure 2a and Table 1); hence, the main process is the shoreline erosion (Figures 10b and 11b). On the contrary, the two other seismic quiescence periods appear incapable of producing significant changes on the accretion/erosion equilibrium.

It is worth mentioning that both the fully automatic DSAS method and the semiautomatic method gave the same results (Table 5). The specific result is in accordance with the results derived in a previous study for a smooth shoreline in the Tyrrhenian Sea, Italy [84]. Both methodologies can be applied to vectorized coastlines independently from their originality (raster basemap used for digitizing the coastline). DSAS was, in the past, used with shorelines digitized from aerial photographs and VHR satellite images [44,45,47,48] or medium resolution satellite data [7,11,12,86]. The main concept in all these studies is that the raster data should have a similar spatial resolution and very accurate georeferencing. Those two prerequisites have been fully accomplished in the current study, as presented in Tables 2 and 4.

Particularly important for understanding the phenomenon of erosion and/or accretion is the last period of monitoring with UAV in the north part of the Egremni beach. During this monitoring, we mapped extensive earthquake-induced landslides and prevalence of erosion. These highly accurate UAV data indicate beyond any doubt that, for a better understanding of the mass wasting after each earthquake, further remote sensing data acquisition and analysis is needed in order to gain knowledge about the exact time that a strong earthquake effect stops its impact as a sediment feeder for the affected shoreline. The basic advantages of the UAV SfM photogrammetry (low cost, flexibility, and high accuracy) proved the feasibility of the specific technology in coastal change mapping. The same conclusion was derived in a recent study [92] using UAV to monitor wind- and wave-driven morphological changes on a beach-dune at Truc-Vert in southwest France.

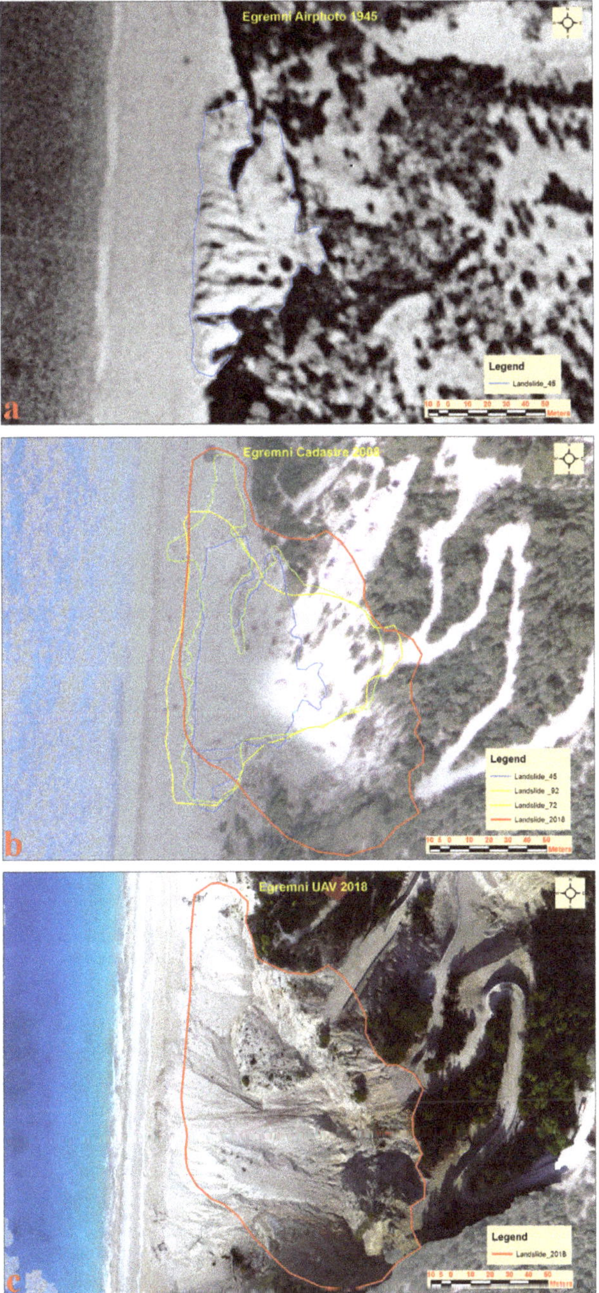

Figure 12. The diachronic evolution of one of the landslides in the overhanging cliff at Egremni beach: From 1945 to 2018, the extent of the landslide almost doubled, offering waste material to Egremni beach. (**a**) The landslide extent as mapped from the 1945 orthophoto. (**b**) Multicolor polygons represent the landslide extent at different dates. The background image is from the 2008 Cadastre orthophoto. (**c**) The landslide extent as mapped from the 2018 UAV orthophoto.

6. Conclusions

The photogrammetric processing of diverse remote sensing data (air photographs, satellite images, and drone-acquired footage) and further analysis in a GIS environment indicates that the southwest shoreline of Lefkada Island is under dynamic equilibrium. This equilibrium is strongly controlled by geological parameters such as the subsidence of the studied shoreline during co-seismic deformation and mass wasting. The following observations can be noted:

1. Headlands appear to control accretion and/or erosion along the sweeping beaches of SW Lefkada Island.
2. Periods of strong seismicity predate shoreline accretion, while periods of relatively long seismic quiescence, in the order of 30 years, are related to erosion.
3. Periods of strong earthquakes give rise to the change of the former prevailed condition. Areas of erosion change to accretion areas and vice versa.
4. Mass wasting is dramatically related to the first years after earthquakes, and sediment dispersal signifies the long-term phenomenon of sediment drift along and across the shoreline.
5. Both famous beaches will maintain their sediment budget, and an analogous evolution will be sustained as long as strong earthquakes occur in the future.

Furthermore, for the first time, it was demonstrated that photogrammetric processing is appropriate for declassified satellite imagery provided by USGS. In order to map the diachronic evolution of a coastline, diverse remote sensing data can be combined. UAVs can be successfully combined with other remote sensing data and have proven to be a very cheap, accurate, and flexible method for coastline monitoring.

Author Contributions: Conceptualization, Konstantinos Nikolakopoulos; methodology, Konstantinos Nikolakopoulos and Vasiliki Zygouri; software, Konstantinos Nikolakopoulos, Aggeliki Kyriou, Vasiliki Zygouri, and Dionysios Apostolopoulos; validation, Konstantinos Nikolakopoulos, Aggeliki Kyriou, and Vasiliki Zygouri; field work, Konstantinos Nikolakopoulos, Ioannis Koukouvelas, Vasiliki Zygouri, and Aggeliki Kyriou; data curation, Aggeliki Kyriou, Vasiliki Zygouri, and Dionysios Apostolopoulos; writing—original draft preparation, Konstantinos Nikolakopoulos, Ioannis Koukouvelas, and Vasiliki Zygouri; writing—review and editing, Konstantinos Nikolakopoulos, Ioannis Koukouvelas, and Vasiliki Zygouri; project administration, Konstantinos Nikolakopoulos and Ioannis Koukouvelas.

Funding: This research received no external funding.

Acknowledgments: The authors would like to thank the "EnCeladus" Hellenic Supersite (http://www.earthobservations.org/gsnl.php) for the Pleiades data provision and the National Greek Cadastre and Mapping Agency for the orthomosaics of 1945 and 2008 offer in the frame of the current study.

Conflicts of Interest: The authors declare no conflict of interest.

References

1. Lavalle, C.; Rocha, G.C.; Baranzelli, C.; Batista e Silva, F. *Coastal Zones Policy Alternatives Impacts on European Coastal Zones 2000–2050*; JRC-IES: European Union's policies: Luxembourg, Luxembourg. 2011. Available online: http://www.jrc.ec.europa.eu/ (accessed on 29 September 2019).
2. Boak Elizabeth, H.; Ian, L. Turner Shoreline Definition and Detection: A Review. *J. Coast. Res.* **2005**, *21*, 688–703. [CrossRef]
3. Feagin, R.A.; Barbier, E.B.; Koch, E.W.; Silliman, B.R.; Hacker, S.D.; Wolanski, E.; Primavera, J.; Granek, E.F.; Polasky, S.; Aswani, S.; et al. Vegetation's role in coastal protection. *Science* **2008**, *320*, 176–177. [CrossRef]
4. Bird, E.C.E. *Beach Management*; John Wiley & Sons: Chichester, UK, 1996.
5. Luijendijk, A.; Hagenaars, G.; Ranasinghe, R.; Baart, F.; Donchyts, G.; Aarninkhof, S. The state of the World's Beaches. *Sci. Rep.* **2018**, *8*, 6641. [CrossRef]
6. Robert, D.; Bruce, H.; Jeffrey, H. A new photogrammetric method for determining shoreline erosion. *Coast. Eng.* **1978**, *2*, 21–39. [CrossRef]

7.   Tigny, V.; Ozer, A.; De Falco, G.; Baroli, M.; Djenidi, S. Relationship between the Evolution of the Shoreline and the Posidonia oceanica Meadow Limit in a Sardinian Coastal Zone. *J. Coast. Res.* **2007**, *23*, 787–793. [CrossRef]

8.   Pasqualini, V.; Pergent-Martini, C.; Clabaut, P.; Marteel, H.; Petgent, G. Integration of Aerial Remote Sensing, Photogrammetry, and GIs Technologies in Seagrass Mapping. *Photograrnmetric Eng. Remote Sens.* **2001**, *67*, 99–105.

9.   Esmail, M.; Mahmod, W.E.; Fath, H. Assessment and prediction of shoreline change using multi-temporal satellite images and statistics: Case study of Damietta coast, Egypt. *Appl. Ocean. Res.* **2019**, *82*, 274–282. [CrossRef]

10. Wenyu, L.; Peng, G. Continuous monitoring of coastline dynamics in western Florida with a 30-Year time series of Landsat imagery. *Remote Sens. Environ.* **2016**, *179*, 196–209. [CrossRef]

11. Salghuna, N.N.; Aravind Bharathvaj, S. Shoreline Change Analysis for Northern Part of the Coromandel Coast. *Aquat. Procedia* **2015**, *4*, 317–324. [CrossRef]

12. Wang, X.; Liu, Y.; Ling, F.; Liu, Y.; Fang, F. Spatio-Temporal Change Detection of Ningbo Coastline Using Landsat Time-Series Images during 1976–2015. *ISPRS Int. J. Geo-Inf.* **2017**, *6*, 68. [CrossRef]

13. Behling, R.; Robert, M.; Sabine, C. Spatiotemporal shoreline dynamics of Namibian coastal lagoons derived by a dense remote sensing time series approach. *Int. J. Appl. Earth Obs. Geoinf.* **2018**, *68*, 262–271. [CrossRef]

14. Thior, M.; Dièye, T.; Sané, T.; El Hadj, B.D.; Sy, O.; Cissokho, D.; Demba Ba, B.; Descroix, L. Coastline dynamics of the northern Lower Casamance (Senegal) and southern Gambia littoral from 1968 to 2017. *J. Afr. Earth Sci.* **2019**, *160*, 103611. [CrossRef]

15. Di, K.; Wang, J.; Ma, R.; Li, R. Automatic shoreline extraction from high-resolution Ikonos satellite imagery. In Proceedings of the Annual ASPRS Conference, Anchorage, AK, USA, 5–9 May 2003.

16. Sekovski, I.; Stecchi, F.; Mancini, F.; Del Rio, L. Image classification methods applied to shoreline extraction on very High-Resolution multispectral imagery. *Int. J. Remote Sens.* **2014**, *35*, 3556–3578. [CrossRef]

17. Nikolakopoulos Konstantinos, G.; Kozarski, D.; Kogkas, S. Coastal areas mapping using UAV photogrammetry. In Proceedings of the SPIE 10428, Earth Resources and Environmental Remote Sensing/GIS Applications VIII, 104280O, Warsaw, Poland, 5 October 2017. [CrossRef]

18. Kaliraj, S.; Chandrasekar, N.; Ramachandran, K.K. Mapping of coastal landforms and volumetric change analysis in the south west coast of Kanyakumari, South India using remote sensing and GIS techniques. *Egypt. J. Remote Sens. Space Sci.* **2016**, *20*, 265–282. [CrossRef]

19. Maglione, P.; Parente, C.; Vallario, A. High resolution satellite images to reconstruct recent evolution of domitian coastline. *Am. J. Appl. Sci.* **2015**, *12*, 506–515. [CrossRef]

20. Görmüş, K.S.; Kutoğlu, Ş.H.; Şeker, D.Z.; Özölçer, I.H.; Oruç, M.; Aksoy, B. Temporal analysis of coastal erosion in Turkey: A case study Karasu coastal region. *J. Coast. Conserv.* **2014**, *18*, 399–414. [CrossRef]

21. Kim, I.-H.; Lee, H.-S.; Song, D.-S. Time series analysis of shoreline changes in Gonghyunjin and Songjiho Beaches, South Korea using aerial photographs and remotely sensed imagery. *J. Coast. Res.* **2013**, *2*, 1415–1420. [CrossRef]

22. Schweitzer, P.N. *Modern Average Global Sea-Surface Temperature*; US Geological Survey Digital Data Series, USGS Numbered Series, Data Series no. 10; U.S. Dept. of the Interior, U.S. Geological Survey: Reston, VA, USA, 1993.

23. Ioannis, K.; Nikolakopoulos, K. Use of Landsat TM images for the detection of water outflows in the coastal area of south Attiki Peninsula, Greece. In Proceedings of the SPIE 4881, Sensors, Systems, and Next-Generation Satellites VI, Crete, Greece, 8 April 2003. [CrossRef]

24. Garrity, C. *Passive Microwave Remote Sensing of Snow Covered Floating Ice during Spring Conditions in the Arctic and Antarctic*; Diss York Univ.: Toronto, ON, Canada, 1991; p. 348.

25. Massom, R.A.; Comiso, J.C.; Worby, A.P.; Lytle, V.I.; Stock, L. Regional classes of sea ice cover in the East Antarctic Pack observed from satellite and in situ data during a winter time period. *Remote Sens. Environ.* **1999**, *68*, 61–76. [CrossRef]

26. Howarth, P.J.; Alfoeldi, T.T.; Laframboise, P.; Munday, J.C.; Thompson, K.P.P.; Tomlins, G.F.; Wickware, G.M. Landsat for monitoring hydrologic and coastal change in Canada. In *Landsat for Monitoring the Changing Geography of Canada. Can. Cent.*; Thompson, M.D., Ed.; Remote Sensing, Energy, Mines and Resour.: Ottawa, ON, Canada, 1982; pp. 7–40.

27. Hanslow, D.J.; Clout, B.; Evans, P.; Coates, B. Islands; Economy, Society and Environment New Zealand Geographical Society Conference Series. *Monit. Coast. Chang. Using Photogramm.* **1997**, *19*, 422–426.

28. Toure, S.; Diop, O.; Kpalma, K.; Maiga, A.S. Shoreline Detection using Optical Remote Sensing: A Review. *ISPRS Int. J. Geo-Inf.* **2019**, *8*, 75. [CrossRef]

29. Chikhradze, N.; Henriques, R.; Elashvili, M.; Kirkitadze, G.; Janelidze, Z.; Bolashvili, N.; Lominadze, G. Close Range Photogrammetry in the Survey of the Coastal Area Geoecological Conditions (on the Example of Portugal). *Earth Sci.* **2015**, *4*, 535–540. [CrossRef]

30. Klemas, V.V. Coastal and Environmental Remote Sensing from Unmanned Aerial Vehicles: An Overview. *J. Coast. Res.* **2015**, *315*, 1260–1267. [CrossRef]

31. Marcaccio, J.V.; Markle, C.E.; Chow-Fraser, P. Unmanned aerial vehicles produce High-Resolution, Seasonally-Relevant imagery for classifying wetland vegetation. *Int. Arch. Photogramm. Remote Sens. Spat. Inf. Sci.-ISPRS Arch.* **2015**, *40*, 249–256. [CrossRef]

32. Papakonstantinou, A.; Topouzelis, K.; Pavlogeorgatos, G. Coastline Zones Identification and 3D Coastal Mapping Using UAV Spatial Data. *ISPRS Int. J. Geo-Inf.* **2016**, *5*, 75. [CrossRef]

33. Drummond, C.D.; Harley, M.D.; Turner, I.L.; Matheen, A.N.A.; Glamore, W.C. UAV applications to coastal engineering. In Proceedings of the Aust. Coasts Ports 2015 Conf., Auckland, New Zealand, 15–18 September 2015; pp. 0–6.

34. Bay, N.; Park, N. UAV Monitoring of Dune Dynamics-Anna Bay Entrance; Stockton Bight. In Proceedings of the NSW Coastal Conference, Coffs Harbour, Australia, 9–11 November 2016; pp. 1–22.

35. Amos, D.A.N. Monitoring Mixed Sand and Gravel Beaches Using Unmanned Aerial Systems. In Proceedings of the Coast. Sediments 2015, San Diego, CA, USA, 11–15 May 2015; pp. 1–13.

36. Turner, I.L.; Harley, M.D.; Drummond, C.D. UAVs for coastal surveying. *Coast. Eng.* **2016**, *114*, 19–24. [CrossRef]

37. Alain, M.; Becker, M.; Benveniste, J.; Cazenave, A. White Paper on Monitoring the evolution of coastal zones under various forcing factors using Space-Based observing systems. In Proceedings of the International Space Science Institute (ISSI) Forum, Bern, Switzerland, 11–12 October 2016; pp. 1–38.

38. Jamie, G.; Barlow, J.; Moore, R. Detection and analysis of mass wasting events in chalk sea cliffs using UAV photogrammetry. *Eng. Geol.* **2019**, *250*, 101–112. [CrossRef]

39. Chen, B.; Yang, Y.; Wen, H.; Ruan, H.; Zhou, Z.; Luo, K.; Zhong, F. High-Resolution monitoring of beach topography and its change using unmanned aerial vehicle imagery. *Ocean. Coast. Manag.* **2018**, *160*, 103–116. [CrossRef]

40. Valderrama-Landeros, L.; Flores-de-Santiago, F. Assessing coastal erosion and accretion trends along two contrasting subtropical rivers based on remote sensing data. *Ocean. Coast. Manag.* **2019**, *169*, 58–67. [CrossRef]

41. Pauline, L.; Jaud, M.; Grandjean, P.; Ammann, J.; Costa, S.; Maquaire, O.; Davidson, R.; Le Dantec, N.; Delacourt, C. Examining High-Resolution survey methods for monitoring cliff erosion at an operational scale. *GIScience Remote. Sens.* **2018**, *55*, 457–476. [CrossRef]

42. Pantanahiran, W. Using remote sensing data for calculating the coastal erosion in southern Thailand. *Int. Arch. Photogramm. Remote Sens. Spat. Inf. Sci.* **2019**, *XLII-3/W7*, 51–56. [CrossRef]

43. Prasita, V.D. Determination of Shoreline Changes from 2002 to 2014 in the Mangrove Conservation Areas of Pamurbaya Using GIS. *Procedia Earth Planet. Sci.* **2015**, *14*, 25–32. [CrossRef]

44. Saci, K.; Boutiba, M.; Guendouz, M. Mohamed Said Guettouche, DalilaKhelfani, Detection and analysis of shoreline changes using geospatial tools and automatic computation: Case of Jijelian sandy coast (East Algeria). *Ocean. Coast. Manag.* **2016**, *132*, 46–58. [CrossRef]

45. Mann, T.; Westphal, H. Assessing Long-Term Changes in the Beach Width of Reef Islands Based on Temporally Fragmented Remote Sensing Data. *Remote Sens.* **2014**, *6*, 6961–6987. [CrossRef]

46. Helene, B.; French, J. Understanding coastal change using shoreline trend analysis supported by cluster-based segmentation. *Geomorphology* **2017**, *282*, 131–149.

47. Ford, M. Shoreline changes interpreted from Multi-Temporal aerial photographs and high resolution satellite images: Wotje Atoll, Marshall Islands. *Remote Sens. Environ.* **2013**, *135*, 130–140. [CrossRef]

48. Ford, M.R.; Kench, P.S. Multi-decadal shoreline changes in response to sea level rise in the Marshall Islands. *Anthropocene* **2015**, *11*, 14–24. [CrossRef]

49. Salim, F.Z.; El Habti, M.Y.; Hamman, L.-H.K.B.; Raissouni, A.; El Arrim, A. Application of a Geomatics Approach for the Diachronic Study of the Meditterannean Coastline Case of Tangier Bay. *Int. J. Geosci.* **2018**, *9*, 320–336. [CrossRef]

50. Poli, D.; Remondino, F.; Angiuli, E.; Agugiaro, G. Radiometric and geometric evaluation of GeoEye-1, WorldView-2 and Pleiades-1A stereo images for 3D information extraction. *ISPRS J. Photogramm. Remote Sens.* **2015**, *100*, 35–47. [CrossRef]

51. Perko, R.; Raggam, H.; Gutjahr, K.H.; Schardt, M. Advanced DTM generation from Very High-Resolution Satellite stereo images. *ISPRS Ann. Photogramm. Remote Sens. Spat. Inf. Sci.* **2015**, *II-3/W4*, 165–172. [CrossRef]

52. Bagnardi, M.; Gonzalez, P.J.; Hooper, A. High-Resolution digital elevation model from Tri-Stereo Pleiades-1 satellite imagery for lava flow volume estimates at Fogo Volcano. *Geophys. Res. Lett.* **2016**, *43*, 6267–6275. [CrossRef]

53. Berthier, E.; Vincent, C.; Magnússon, E.; Gunnlaugsson, Á.Þ.; Pitte, P.; Le Meur, E.; Masiokas, M.; Ruiz, L.; Pálsson, F.; Belart, J.M.C.; et al. Glacier topography and elevation changes derived from Pléiades Sub-Meter stereo images. *Cryosphere* **2014**, *8*, 2275–2291. [CrossRef]

54. Panagiotakis, E.; Chrysoulakis, N.; Charalampopoulou, V.; Poursanidis, D. Validation of Pleiades Tri-Stereo DSM in Urban Areas. *ISPRS Int. J. Geo-Inf.* **2018**, *7*, 118. [CrossRef]

55. Kokkalas, S.; Xypolias, P.; Koukouvelas, I.; Doutsos, T. Postcollisionalcontractional and extensional deformation in the Aegean region. *Spec. Pap. Geol. Soc. Am.* **2006**, *409*, 97–123. [CrossRef]

56. Kahle, H.-G.; Müller, M.V.; Geiger, A.; Danuser, G.; Mueller, S.; Veis, G.; Billiris, H.; Paradissis, D. The strain field in the northwestern Greece and the Ionian Islands: Results inferred from GPS measurements. *Tectonophysics* **1995**, *249*, 41–52. [CrossRef]

57. Louvari, E.; Kiratzi, A.A.; Papazachos, B.C. The Cephalonia Transform Fault and its extension to western Lefkada island. *Tectonophysics* **1999**, *308*, 223–236. [CrossRef]

58. Papazachos, B.C.; Papazachou, C.C. *The Earthquakes of Greece*; Ziti Publication Co.: Thessaloniki, Greece, 1997.

59. Papadopoulos, G.A.; Karastathis, V.K.; Koukouvelas, I.; Sachpazi, M.; Baskoutas, I.; Chouliaras, G.; Agalos, A.; Daskalaki, E.; Minadakis, G.; Moshou, A.; et al. The Cephalonia, Ionian Sea (Greece), sequence of strong earthquakes of January–February 2014: A first report. *Res. Geophys.* **2014**, *4*, 5441. [CrossRef]

60. Fokaefs, A.; Papadopoulos, G.A. Historical earthquakes in the region of Lefkada Island, Ionian Sea–Estimation of magnitudes from epicentral intensities. *Bull. Geol. Soc. Greece* **2004**, *36*, 1389–1395. [CrossRef]

61. Papadopoulos, G.; Karastathis, V.; Ganas, A.; Pavlides, S.; Fokaefs, A.; Orfanogiannaki, K. The LefkadaIonian sea (Greece), Shock (Mw6.2) of 14 August 2003: Evidence for the characteristic earthquake from seismicity and ground failures. *Earth Planets Space* **2003**, *55*, 713–718. [CrossRef]

62. Sokos, E.; Zahradnik, J.; Gallovic, F.; Serpetsidaki, A.; Plicka, V.; Kiratzi, A. Asperity break after 12 years: The Mw6.4 2015 Lefkada(Greece) earthquake. *Geophys. Res. Lett.* **2016**, *43*, 6137–6145. [CrossRef]

63. Zygouri, V.; Koukouvelas, I. Techniques of rockfall inventory in earthquake prone rock slopes. *Bull. Seism. Soc. Greece* **2016**, *50*, 1756–1765. [CrossRef]

64. Papadopoulos, G.; Agalos, A.; Biocchini, G.M.; Chousianitis, K.; Karastathis, V.; Triantafyllou, I.; Kontoes, C.; Papoutsis, I.; Svigkas, N.; Koukouvelas, I.; et al. The Mw6.5 earthquake of 17 November 2015 in Lefkada Island and the seismotectonics in the Cephalonia Transform Fault (Ionian Sea, Greece). *Geophys. Res. Abstr.* **2016**, *18*, EGU2016-9041-1.

65. Tsangaratos, P.; Loupasakis, C.; Nikolakopoulos, K.; Angelitsa, V.; Ilia, I. Developing a landslide susceptibility map based on remote sensing, fuzzy logic and expert knowledge of the Island of Lefkada, Greece. *Environ. Earth Sci.* **2018**, *77*, 363. [CrossRef]

66. Cushing, M. Evolution structurale de la marge nord ouest hellénique dans l' île de Levkas et ses environs (Grèce nord occidentale). In *Thèse 3me Cycle*; Univ. de Paris-Sud: Paris, France, 1985.

67. Rondoyanni, T.; Mettos, A.; Paschos, P.; Georgiou, C. *Neotectonic Map of Greece, Scale 1:100.000*; Lefkada sheet. I.G.M.E.: Athens, Greece, 2007.

68. Papazachos, B.C.; Karakostas, V.G.; Papazachos, C.B.; Scordilis, E.M. The geometry of the Wadati-Benioff zone and lithospheric kinematics in the Hellenic arc. *Tectonophysics* **2000**, *319*, 275–300. [CrossRef]

69. Karakostas, V.; Papadimitriou, E.; Patias, P.; Georgiadis, C. Coastal deformation in Lefkada Island associated with strong earthquake occurrence. *Bolletino Geofis. Appl.* **2019**, *60*, 1–16.

70. Declassified Intelligence Satellite Photographs. FS 2008-3054. Geological Survey (U.S.); 2008. Available online: https://pubs.usgs.gov/fs/2008/3054/pdf/fs2008-3054.pdf (accessed on 29 September 2019).

71. Nikolakopoulos Konstantinos, G.; Kavoura, K.; Sabatakakis, N.; Vaiopoulos, A.D. Fusion of declassified airphotos and Landsat MSS data for old landslides detection. In Proceedings of the SPIE 9245, Earth Resources and Environmental Remote Sensing/GIS Applications V, 92450E, Amsterdam, The Netherlands, 23 October 2014. [CrossRef]

72. Schenk, T.; Csatho, B.; Shin, S.W. Rigorous panoramic camera model for DISP Imagery. In Proceedings of the Joint ISPRS/EARSeL Workshop 'High Resolution Mapping from Space 2003', Hannover, Germany, 6–8 October 2003.

73. Sohn, H.G.; Kim, G.H.; Yom, J.H. Mathematical modelling of historical reconnaissance Corona KH-4B Imagery. *Photogramm. Rec.* **2004**, *19*, 51–66. [CrossRef]

74. Centre National d'Etudes Spatiales (CNES). 2016. Available online: https://pleiades.cnes.fr/en/PLEIADES/index.htm (accessed on 10 May 2018).

75. Nikolakopoulos, K.G.; Soura, K.; Koukouvelas, I.K. Argyropoulos, N.G. UAV vs classical aerial photogrammetry for archaeological studies. *J. Archaeol. Sci. Rep.* **2017**, *14*, 758–773. [CrossRef]

76. Casana, J.; Cothren, J. Stereo analysis, DEM extraction and orthorectification of CORONA satellite imagery: Archaeological applications from the Near East. *Antiquity* **2008**, *82*, 732–749. [CrossRef]

77. Galiatsatos, N.; Donoghue, D.N.M.; Philip, G. High resolution elevation data derived from stereoscopic CORONA imagery with minimal ground control: An approach using IKONOS and SRTM data. *Photogramm. Eng. Remote Sens.* **2008**, *74*, 1093–1106. [CrossRef]

78. Altmaier, A.; Kany, C. Digital surface model generation from CORONA satellite images. *ISPRS J. Photogramm. Remote Sens.* **2002**, *56*, 221–235. [CrossRef]

79. Hamandawana, H.; Eckardt, F.; Ringrose, S. Proposed methodology for georeferencing and mosaicking Corona photographs. *Int. J. Remote Sens.* **2007**, *28*, 5–22. [CrossRef]

80. Konstantinos, N.; Pavlopoulos, K.; Chalkias, C.; Manou, D. Monitoring the urban expansion of Athens using remote sensing and GIS techniques in the last 35 years. In Proceedings of the SPIE 5983, Remote Sensing for Environmental Monitoring, GIS Applications, and Geology V, 598309, Bruges, Belgium, 31 October 2005. [CrossRef]

81. Nikolakopoulos, K.; Kavoura, K.; Depountis, N.; Kyriou, A.; Argyropoulos, N.; Koukouvelas, I.; Sabatakakis, N. Preliminary results from active landslide monitoring using multidisciplinary surveys. *Eur. J. Remote Sens.* **2017**, *50*, 280–299. [CrossRef]

82. Nikolakopoulos, K.G.; Lampropoulou, P.; Fakiris, E.; Sardelianos, D.; Papatheodorou, G. Synergistic Use of UAV and USV Data and Petrographic Analyses for the Investigation of Beachrock Formations: A Case Study from Syros Island, Aegean Sea, Greece. *Minerals* **2018**, *8*, 534. [CrossRef]

83. Westoby, M.; Brasington, J.; Glasser, N.F.; Hambrey, M.J.; Reyonds, M.J. 'Structure–from–Motion' photogrammetry: A low cost, effective tool for geoscience applications. *Geomorphology* **2012**, *179*, 300–314. [CrossRef]

84. Thieler, E.R.; Danforth, W.W. Historical shoreline mapping (II): Application of the digital shoreline mapping and analysis systems (DSMS/DSAS) to shoreline change mapping in Puerto Rico. *J. Coast. Res.* **1994**, *10*, 600–620.

85. Anfuso, G.; Bowman, D.; Danese, C.; Pranzini, E. Transect based analysis versus area based analysis to quantify shoreline displacement: Spatial resolution issues. *Environ. Monit. Assess.* **2016**, *188*, 568. [CrossRef] [PubMed]

86. RajuAedla, G.S.; Dwarakish, D. Venkat Reddy, Automatic Shoreline Detection and Change Detection Analysis of Netravati-GurpurRivermouth Using Histogram Equalization and Adaptive Thresholding Techniques. *Aquat. Procedia* **2015**, *4*, 563–570. [CrossRef]

87. Ghionis, G.; Poulos, S.; Verykiou, E.; Karditsa, A.; Alexandrakis, G.; Andris, P. The Impact of an Extreme Storm Event on the Barrier Beach of the Lefkada Lagoon, NE Ionian Sea (Greece). *Mediterr. Mar. Sci.* **2015**, *16*, 562–572. [CrossRef]

88. Athanassoulis, G.A.; Skarsoulis, E.K. *Wind and Wave Atlas of North-East Mediterranean Sea. Laboratory of Nautical and Marine Hydrodynamics*; Hellenic Navy General Staff, Hellenic Army Navy, NTUA: Athens, Greece, 1992; p. 191.

ISPRS Int. J. Geo-Inf. **2019**, *8*, 489

89. Padró, J.-C.; Muñoz, F.-J.; Planas, J.; Pons, X. Comparison of four UAV georeferencing methods for environmental monitoring purposes focusing on the combined use with airborne and satellite remote sensing platforms. *Int. J. Appl. Earth Obs. Geoinf.* **2019**, *75*, 130–140. [CrossRef]
90. Rabah, M.; Basiouny, M.; Ghanem, E.; Elhadary, A. Using RTK and VRS in direct geo-referencing of the UAV imagery. *NRIAG J. Astron. Geophys.* **2018**, *7*, 220–226. [CrossRef]
91. Ganas, A.; Elias, P.; Bozionelos, G.; Papathanassiou, G.; Avallone, A.; Papastergios, A.; Valkaniotis, S.; Parcharidis, I.; Briole, P. Coseismic deformation, field observations and seismic fault of the 17 November 2015 Mw6.5, lefkada Island, Greece earthquake. *Tectonophysics* **2016**, *687*, 210–222. [CrossRef]
92. Laporte-Fauret, Q.; Marieu, V.; Castelle, B.; Michalet, R.; Bujan, S.; Rosebery, D. Low-Cost UAV for High-Resolution and Large-Scale Coastal Dune Change Monitoring Using Photogrammetry. *J. Mar. Sci. Eng.* **2019**, *7*, 63. [CrossRef]

International Journal of
Geo-Information

Article

Mapping Canopy Heights of Poplar Plantations in Plain Areas Using ZY3-02 Stereo and Multispectral Data

Mingbo Liu [1,2], **Chunxiang Cao** [1,*], **Wei Chen** [1] **and Xuejun Wang** [3]

1   Institute of Remote Sensing and Digital Earth, Chinese Academy of Sciences, Beijing 100101, China; liumb@radi.ac.cn (M.L.); chenwei@radi.ac.cn (W.C.)
2   University of Chinese Academy of Sciences, Beijing 100049, China
3   Academy of Forest Inventory and Planning, State Forestry Administration, Beijing 100714, China; wangxuejun320@126.com
*   Correspondence: caocx@radi.ac.cn; Tel.: +86-010-6483-6205

Received: 17 January 2019; Accepted: 23 February 2019; Published: 27 February 2019

Abstract: Forest canopy height plays an important role in forest management and ecosystem modeling. There are a variety of techniques employed to map forest height using remote sensing data but it is still necessary to explore the use of new data and methods. In this study, we demonstrate an approach for mapping canopy heights of poplar plantations in plain areas through a combination of stereo and multispectral data from China's latest civilian stereo mapping satellite ZY3-02. First, a digital surface model (DSM) was extracted using photogrammetry methods. Then, canopy samples and ground samples were selected through manual interpretation. Canopy height samples were obtained by calculating the DSM elevation differences between the canopy samples and ground samples. A regression model was used to correlate the reflectance of a ZY3-02 multispectral image with the canopy height samples, in which the red band and green band reflectance were selected as predictors. Finally, the model was extrapolated to the entire study area and a wall-to-wall forest canopy height map was obtained. The validation of the predicted canopy height map reported a coefficient of determination (R^2) of 0.72 and a root mean square error (RMSE) of 1.58 m. This study demonstrates the capacity of ZY3-02 data for mapping the canopy height of pure plantations in plain areas.

Keywords: canopy height; ZY3-02; photogrammetry; poplar plantation; plain area

1. Introduction

Forest canopy height is a critical parameter for forest management and ecosystem modeling. Field measurements are essential but labor intensive and costly, especially in extensive and remote areas [1,2]. A variety of methods for mapping forest height using remote sensing data have been developed in recent years: the light detection and ranging (LiDAR) technique uses point cloud or waveform data from laser pulses to detect the vertical structure of a forest. Distance is determined by measuring the time taken for the laser pulse to travel between the sensor and target. In recent years, the airborne laser scanning (ALS) technique has made rapid progress and is now widely used in forestry applications. It has proven to be a reliable LiDAR method which can measure forest height with sub-meter accuracy [3,4]. The Geoscience Laser Altimeter System (GLAS) onboard NASA's Ice, Cloud, and land Elevation Satellite (ICESat) collected data from 2003 to 2009. It offered an unprecedented opportunity for estimating forest height at a global scale [5]. Lefsky and Simard et al. have produced global forest height maps by combining the GLAS data with Moderate Resolution Imaging Spectroradiometer (MODIS) data [6,7]. Recent developments include new terrain correction methods [8] and the combination of GLAS data with Landsat time-series data [9]. It is expected that the application of spaceborne LiDAR in large-scale forest height mapping will be further developed when data from the ICESat-2 [10] and GEDI (Global Ecosystem Dynamics Investigation, a LiDAR system mounted on the International Space Station) [11] become available.

The synthetic aperture radar (SAR) techniques used for forest height mapping include radargrammetry, interferometry SAR (InSAR), and polarimetric interferometric SAR (PolInSAR). A major advantage of SAR is its high temporal resolution, because SAR has the ability to acquire data under various illumination and weather conditions. Radargrammetry is based on SAR stereo images. It has recently received attention because of the emergence of high spatial resolution SAR images [12]. Radargrammetry needs a digital terrain model (DTM) for deriving forest height [13]. InSAR is based on the phase differences between two complex SAR images, which also need a DTM for deriving forest height. In addition, the time interval between the two images should be short to avoid decoherence [14,15]. PolInSAR includes phase difference methods and model-based methods [16,17]. Based on the coherence optimization method, Cloude and Papathanassiou were the first to use PolInSAR in forest height mapping [18]. Yamada et al. have used the ESPRIT method to separate phase centers of the canopy and ground [19]. The random volume over ground (RVoG) model is the most commonly used model among the model-based methods. Papathanassiou et al. have used the six-dimension non-liner iteration method for the inversion of the RVoG model [20]. Cloude and Papathanassiou have proposed the three-stage inversion method, which has improved the accuracy and reduced the complexity of the inversion process [21]. The most commonly used SAR data include the TerraSAR-X (X-band), Radarsat-2 (C-band), and ALOS-2 (L-band). The data from the Biomass (P-band) (due for launch in 2021) and Tandem-L (L-band) (due for launch in 2022) missions will alleviate the shortage of L-band and P-band SAR data suitable for the extraction of forest structural parameters [22,23].

Since the emergence of advanced sensors has improved the radiometric and spatial resolution, and advances in computing technology have made complex algorithms for image matching practical, the application of digital photogrammetry (DP) in forestry has received more and more attention in recent years [24,25]. DP uses a rigorous or rational function model to characterize the geometry of the acquisition system and uses parallax differences to compute the coordinates and elevations of the matched pixels. Airborne DP systems and several spaceborne DP systems, such as the WorldView series, provide sub-meter resolution stereo images from which a high-quality digital surface model (DSM) can be derived [26]. The forest canopy height can be obtained by calculating the difference between the DSM and the ALS-derived or field-surveyed DTM that provides the bare earth elevation [27–29]. Because the cost of space-borne DP data is usually less, it is ideal for repetitive forest canopy height surveys over a wide area. In addition, due to the long history of the photogrammetry technique, there are a considerable number of photographs in several countries. By digitizing these photographs, researchers may be able to study changes in forest height over a long period of time [29].

However, compared with LiDAR-based techniques, few studies have been carried out on mapping forest canopy heights using surging stereo images. This is partly due to the lack of DTMs. One of the objectives of this study was to evaluate an alternative method for obtaining canopy height samples. In this study, we extracted a DSM from the stereo images of the ZY3-02 satellite. Canopy height samples of poplar plantations were derived from the DSM based on manually selected canopy and ground samples. In order to obtain a canopy height map with complete horizontal coverage, we combined canopy height samples with multispectral image. There is theoretical and empirical evidence indicating that both biochemical properties (e.g. water and pigment content) and the structures of forests are the driving forces regarding the response of multispectral reflectance to canopy height [30,31]. For example, higher forests usually have higher chlorophyll content per unit area, resulting in stronger reflections in the green band. At the same time, higher forests usually have more complex structures that create more shadows, causing less light to be reflected into the sensor. The dominant effect depends on the specific spectral band and forest type. Many researchers have utilized the correlation between the reflectance of multispectral images and forest height in their studies [6,9,32]. In this study, we correlated the reflectance of a ZY3-02 multispectral image with canopy height samples through a regression model and produced a wall-to-wall forest canopy height map of the entire study area after extrapolation. Hence, this study is also an evaluation of the potential of ZY3-02 data in forest canopy height mapping. In addition, we propose methods for setting ground control points (GCPs) and extracting the GCPs' elevations from free terrain data.

2. Study Area

Our study area was located on the southern edge of Horqin Sandy Land, eastern Inner Mongolia, China, with a total area of 2500 km^2 (42°17′11″–42°48′44″N, 119°34′6″–120°18′15″E) (Figure 1). The area belongs to the mid-temperate sub-arid zone, with an annual precipitation between 310 and 460 mm. The altitude there ranges from 396 to 855 m. The terrain is relatively flat with an average slope of 3.4°. Most of the forests in the region are poplar plantations, aged between five and 25 years. These plantations are grid- and patch-shaped. They were planted for the Three-North shelter forest program, a large-scale afforestation project in China which began in 1978, and play an important role in blocking sandstorms, preventing soil erosion, and improving the ecological environment.

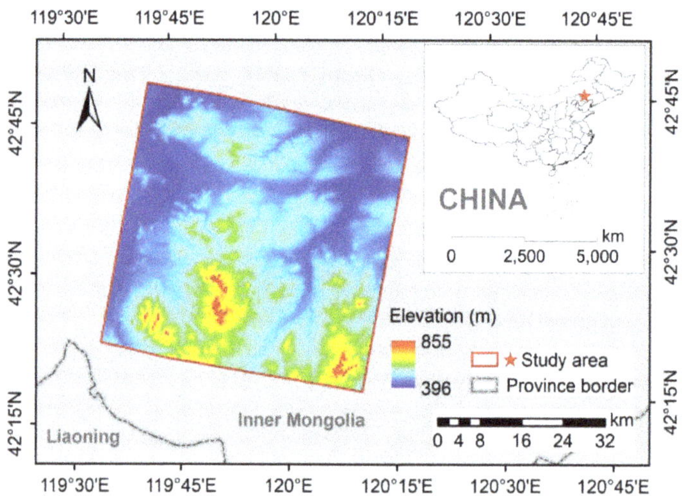

Figure 1. Location and elevation of the study area.

3. Data

3.1. ZY3-02 Data

The ZY3-02 satellite was launched on May 30, 2016, with a design life of five years. It is an upgraded successor to China's first civilian stereo mapping satellite, ZY-3, which was launched in 2012. The ZY3-02 satellite operates in a sun-synchronous near-polar orbit at an altitude of 505 km. It carries a multispectral camera and three panchromatic cameras pointing forward, nadir, and backward. The multispectral camera consists of four channels: blue (450–520 nm), green (520–590 nm), red (630–690 nm), and near infrared (770–890 nm), with a ground sample distance (GSD) of 5.8 m. The forward and backward cameras are arranged at an inclination of ±22° from nadir to realize a base-to-height (B/H) ratio of 0.88. The GSD of the nadir panchromatic camera is 2.1 m. The GSD of the forward and backward panchromatic cameras have been improved from 3.5 m in ZY-3 to 2.7 m. ZY3-02 images have a radiometric resolution of 10 bits and a swath width of 50 km.

According to an assessment made by Xu et al. [33], the planimetric and vertical accuracies of the ZY3-02 sensor-corrected products are better than 2.5 m and 2 m, respectively, with a few GCPs.

The ZY3-02 sensor-corrected products used in this study were acquired on September 18 2017 and were provided by the China Centre for Resources Satellite Data and Application (CRESDA, http://www.cresda.com/CN/). We used the forward/backward stereo images to extract the DSM. The multispectral image was used to classify the landcover and predict forest canopy heights. Both the stereo and multispectral images were able to completely cover the study area. At the time of data acquisition, there was no cloud over the study area.

3.2. SRTMGL1 Data

The Shuttle Radar Topography Mission (SRTM) collected data with a C-band radar interferometry system onboard the space shuttle Endeavour from February 11 to February 22, 2000. The SRTM product provides the elevation of the land between 60° north and 56° south latitude (covering more than 80% of Earth's total landmass) [34]. The SRTM product was first released in 2003 and has been updated several times since. The SRTM global 1 arc second data (SRTMGL1) for China have been available since July 2015 and have a spatial resolution of about 27 m for our study area. The vertical accuracy of SRTMGL1 in plain areas is about 1.9 m according to an assessment made by Hu et al. [35]. We downloaded the void-filled SRTMGL1 Version 3 product from NASA's website (https://earthdata. nasa.gov/). This data was used to determine the elevation of GCPs.

4. Methods

In this study we first extracted a DSM from the ZY3-02 forward/backward stereo images using photogrammetry methods. Then, the canopy samples and ground samples were selected through the manual interpretation of a ZY3-02 multispectral image. The canopy height samples were obtained by calculating the DSM elevation difference for each pair of canopy/ground samples. After that, the canopy height samples were divided into a training set and validation set. A multiple linear regression model was established to correlate the reflectance of the multispectral image with the canopy heights of the training samples. Finally, the model was extrapolated to all forest regions in the study area and a wall-to-wall forest canopy height map was obtained. The map was validated independently by the validation samples (Figure 2).

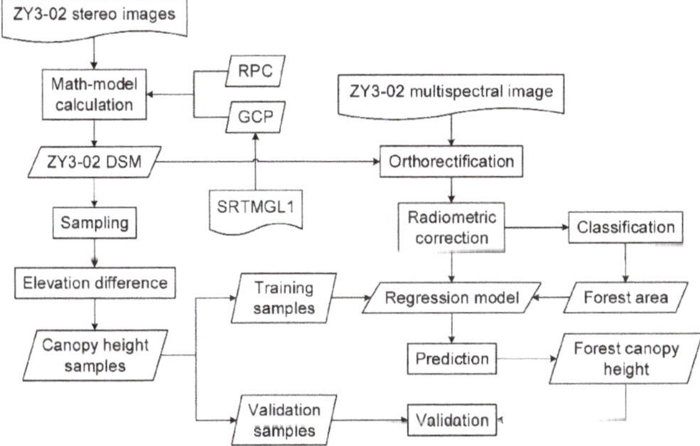

Figure 2. Flow chart of methods. Legend: RPC, rational polynomial coefficients; GCP, ground control point; DSM, digital surface model; SRTMGL1, Shuttle Radar Topography Mission global 1 arc second data.

4.1. DSM Extraction

We extracted DSM using the OrthoEngine module of the PCI Geomatica software (PCI Geomatics Enterprises, Inc., Canada). The module uses the polynomial coefficients, GCPs, and tie points (TIEs) to compute a math model that relates the rows and columns of the matched pixels with ground coordinates and elevations. The polynomial coefficients were provided in the rational polynomial coefficients (RPC) files distributed with the ZY3-02 data. GCPs were set evenly throughout the study area referencing the ZY3-02 multispectral image and ESRI's online World Imagery (Environmental Systems Research Institute, Inc., United States). In this study, we extracted the elevation of GCPs from the SRTMGL1 to minimize fieldwork. However, this brought some difficulties to locating the GCPs. Most of the surface features in the study area were located in valleys, buildings, and at the intersections of forest belts, where elevation changes occur. The values of the SRTMGL1 at these locations may be affected by nearby elevation changes, since the spatial resolution of the SRTMGL1 is about 27 m in the study area. This may introduce bias to the GCPs. To avoid this problem, the GCPs were placed in flat regions and their locations were determined by the intersections of the lines connecting surface features. Figure 3 illustrates the scheme for setting GCPs on ESRI's online World Imagery layer. Figure 4 shows the locations of all 49 GCPs on the true color composite ZY3-02 multispectral image. In the extraction of the DSM, we first rediscovered the four surface features for each GCP in the forward and backward view images, then located the GCPs by connecting the features.

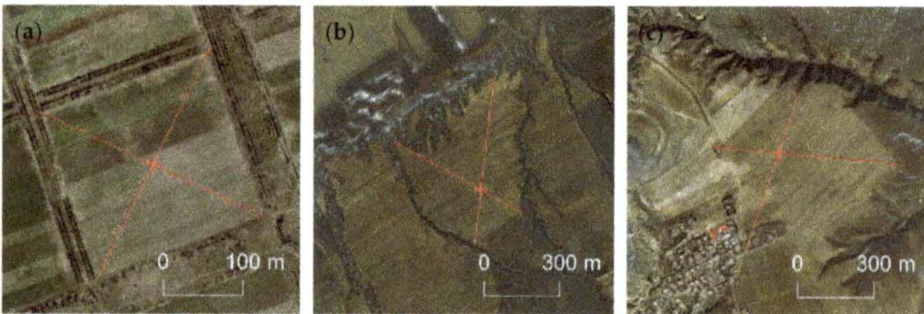

Figure 3. Scheme for setting GCPs on ESRI's online World Imagery layer (Environmental Systems Research Institute, Inc., United States). Subgraphs (**a–c**) show the locations of GCPs determined by intersections of forest belts, ditches in farmlands, and edges of buildings and valleys, respectively.

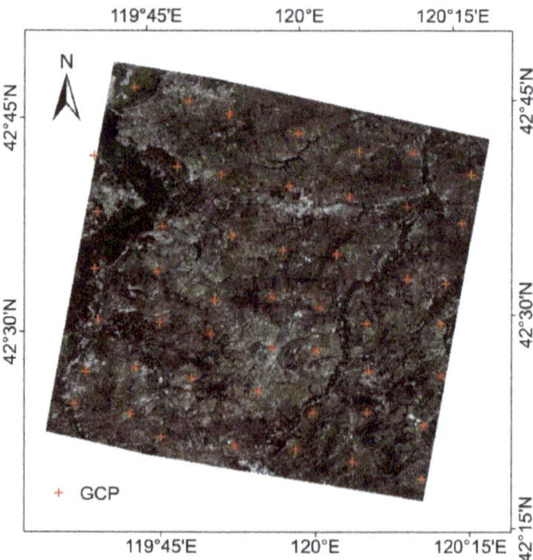

Figure 4. Locations of all 49 GCPs on the true color composite ZY3-02 multispectral image.

In addition, we collected a total of 226 TIEs interactively in the forward and backward view images. The Semi-Global Matching (SGM) method [36] was used to match pixels. In order to avoid the loss of precision, the sampling distance of the output DSM was set to 2 m. In order to preserve all the details, we did not filter the DSM.

4.2. Sample Selection

We selected 51 pairs of canopy/ground samples in flat areas through the manual interpretation of the ZY3-02 multispectral image (Figure 5a,c). Areas affected by shadows and mismatches during the DSM extraction process were excluded from the sample selection, referencing the ZY3-02 DSM displayed with hill shade effect (Figure 5b,d). Canopy samples were set in relatively uniform and closed forests. The elevation undulation of the surrounding bare ground needed to be within 2 m (i.e., the vertical accuracy of ZY3-02) so that the sampled forests were unlikely to be on slopes. The ground samples were set in flat areas without trees, buildings, or ditches and as close as possible to the canopy samples, to make the elevation of the ground samples close to the elevation of the ground beneath the sampled forest canopies. The attributes of the sampling areas are summarized in Table 1. We averaged the DSM in each sampling area and obtained the canopy height samples by calculating the difference of the averaged DSM elevation for each pair of canopy/ground samples. The canopy height samples had a minimum height of 0.5 m, a maximum height of 14.3 m, and an average height of 6.1 m.

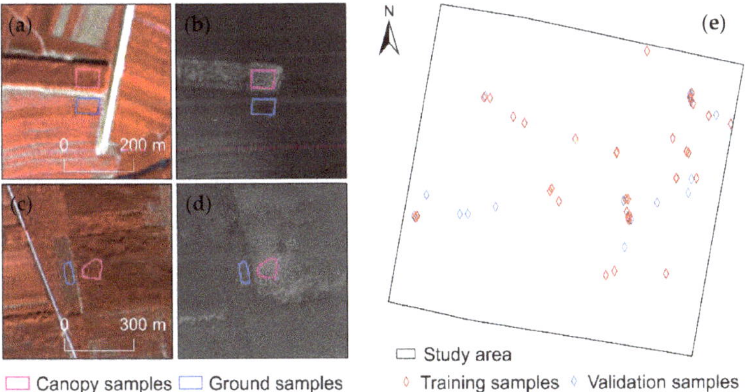

Figure 5. Scheme of sample selection. Subgraphs (**a**–**d**) illustrate the setting of canopy samples and ground samples referencing the false color composite ZY3-02 multispectral image and the ZY3-02 DSM displayed with hill shade effect. Subgraph (**e**) shows the locations of all 51 canopy height samples (which were also the locations of the canopy samples) and marks the training samples and validation samples.

Table 1. Area, DSM pixel counts, and mean DSM elevation of the sampling areas.

	Area (m²)	DSM Pixels	Mean Elevation (m)
Canopy samples	748–22,864	187–5,716	420–614
Ground samples	620–17,288	155–4,322	415–611

4.3. Canopy Height Modeling and Extrapolation

The canopy height samples were randomly divided into a training set and a validation set, with 34 samples in the training set and 17 samples in the validation set (Figure 5e). We first performed geometric correction to the ZY3-02 multispectral image using the ZY3-02 DSM. Then, we performed radiometric correction to obtain the surface reflectance of each band through the radiometric correction workflow in PCI Geomatica software (PCI Geomatics Enterprises, Inc., Canada). The atmospheric condition of the input image was estimated based on the season when the image was captured and the location at which the image was taken. The workflow works with a database of atmospheric correction functions stored in lookup tables for different altitude profiles of pressure, humidity, and aerosol type [37]. We built a multiple linear regression model to correlate the mean surface reflectance of the multispectral image in sampling areas with the canopy heights derived from the ZY3-02 DSM. The modeling was implemented in SPSS software (International Business Machines Corp., United States) and a stepwise method was used to select variables [32].

We classified the ZY3-02 multispectral image using the maximum likelihood supervised classification algorithm. The algorithm assumes that the statistics for each class in each band are normally distributed and calculates the probability that a given pixel belongs to a specific class. Each pixel is assigned to the class that has the highest probability (i.e., the maximum likelihood) based on the probability density functions estimated from the training samples. The classification was implemented through classification workflow using ENVI software (Harris Geospatial Solutions, Inc., United States). Training samples were selected in the multispectral image through manual interpretation. The number of samples for forest, water, bare surfaces, and farmland were 80, 20, 64, and 58, respectively. The classification results were exported as raster files without smoothing or aggregation.

After that, the surface reflectance of the multispectral image in forest regions was input into the regression model and a wall-to-wall forest canopy height map for the forest regions throughout the study area was obtained.

5. Results

Figure 6 shows the comparison of the ZY3-02 DSM and SRTMGL1 at the locations of the GCPs. The coefficient of determination (R^2), mean error (μ), and standard deviation of errors (σ) were 0.9992, 0.04 m, and 1.91 m, respectively.

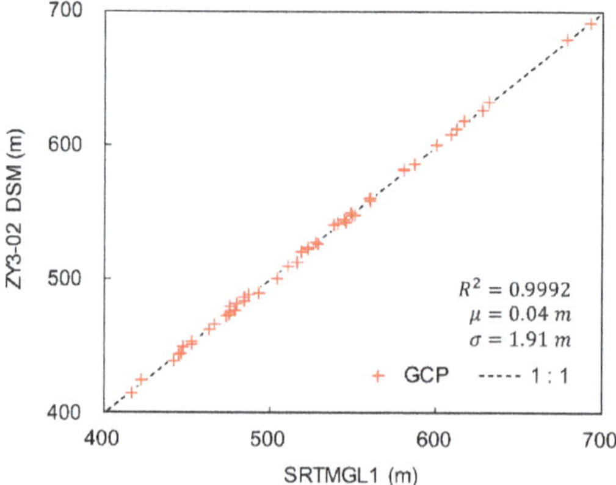

Figure 6. Comparison of the ZY3-02 DSM and SRTMGL1 at the locations of the GCPs.

After the stepwise process, the red band and green band were selected as the predictors of the optimal model. The model was of the form

$$H = 1.3219 - 0.1013 * Red + 0.0713 * Green \tag{1}$$

where H is the forest canopy height, Red is the surface reflectance of the red band, and Green is the surface reflectance of the green band. The model and predictors were statistically significant, with the *p*-values of the F-test (for the model) and t-test (for the predictors) being less than 0.01. The model had an R^2 of 0.67 and a root mean square error (RMSE) of 1.55 m (Figure 7).

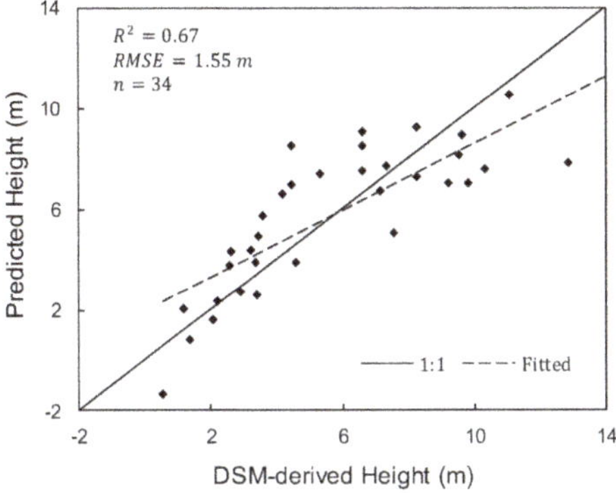

Figure 7. Comparison of the model-predicted heights and the DSM-derived heights for the training samples. Legend: R^2, coefficient of determination; RMSE, root mean square error.

The accuracy of the landcover classification was evaluated by 500 points randomly set in the study area. The reference categories of each point were determined by manual interpretation and were used to validate the results of the supervised classification. The confusion matrix showed an overall accuracy of 84%. The predicted forest canopy height map of the entire study area is shown in Figure 8a. Figure 8b,c show the canopy height maps of two regions corresponding to the subgraphs (a,b) and (c,d) in Figure 5.

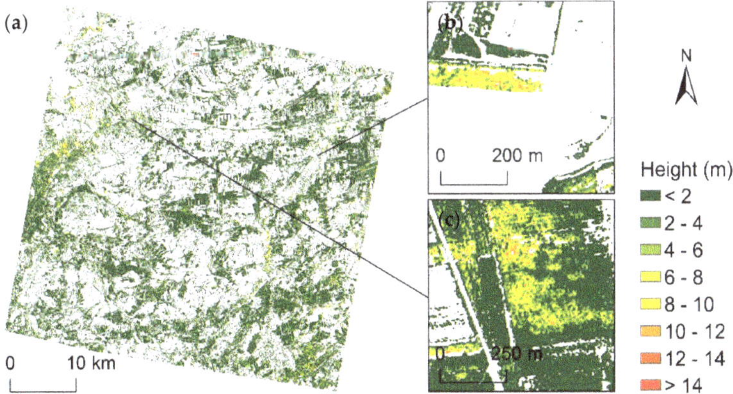

Figure 8. Subgraph (**a**) shows the predicted forest canopy height map of the entire study area. Subgraphs (**b**,**c**) show the canopy height maps of two regions corresponding to the subgraphs (**a**,**b**) and (**c**,**d**) in Figure 5.

To validate the predicted forest canopy height, we calculated the mean heights in areas of the validation samples and compared them with the DSM-derived canopy heights. The validation reported an R^2 of 0.72 and a RMSE of 1.58 m (Figure 9).

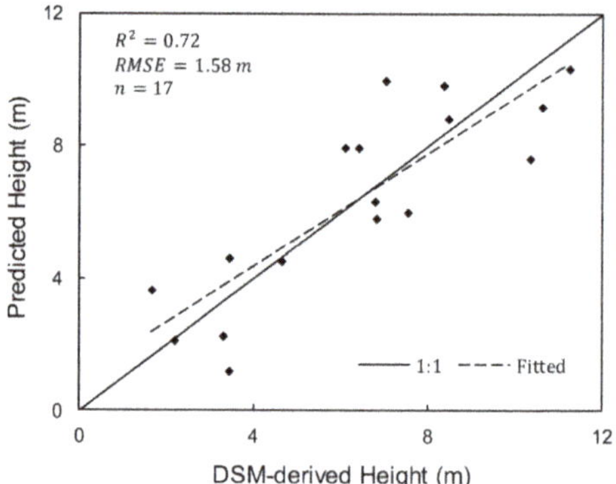

Figure 9. Comparison of the model-predicted heights and the DSM-derived heights for the validation samples.

6. Discussion

The use of the intersections of the lines connecting surface features showed good performance in determining the locations of the GCPs. The standard deviation of 1.91 m in Figure 6 was slightly better than the vertical accuracy of 2 m for ZY3-02 reported by Xu et al. [33]. The canopy samples and ground samples were selected through manual interpretation to ensure the quality of canopy height samples. After selecting the samples, it was necessary to correlate the reflectance of the multispectral image with the DSM-derived canopy heights through a statistical model in order to obtain a wall-to-wall canopy height map. Since the forests in the study area were relatively short, a linear model would have been able to effectively characterize the relationship between the reflectance and canopy height [32]. However, in tall forests (e.g. taller than 30 m), the relationship between reflectance and canopy height may deviate from linear form and a nonlinear model may be required in that case [9]. The reflectance of the red band contributed most in the estimation of canopy height, which was consistent with the study of Hansen et al. [9].

We did not use field data to validate our results. However, the accuracy of the DSM-derived canopy height samples should be consistent with the vertical accuracy of ZY3-02, which is better than 2 m based on the evaluations conducted by Xu et al. [33]. In addition, we used the average DSM elevation in sampling areas to calculate the canopy height samples. This further reduced the likelihood of error and made the accuracy of the DSM-derived canopy heights closer to that of the field measurements [38].

The results in Figure 9 were acceptable considering the planimetric and vertical accuracies of the stereo images and the limited bands of the multispectral image. However, in order to better meet the requirements of forestry applications, further improvements would be necessary in future. Finer and better stereo images are worth trying to obtain, if available. For example, China's Gaofen-7 stereo mapping satellite, due for launch in 2019, provides stereo images with sub-meter spatial resolution and can carry out topographic mapping on a scale of 1:10,000. It would also be helpful to use more spectral bands such as the short-wave near-infrared band, which has a strong response to the water content of canopies. With more bands, various vegetation indices can be used in the modeling, which will help to reduce the effects of illumination conditions and canopy surface undulations. Finally, machine learning algorithms may be useful because they usually do not require assumptions about the form of the model, and this will also be the direction of our future work.

7. Conclusions

In this study, we mapped the canopy heights of poplar plantations in plain areas using ZY3-02 data by combining photogrammetry with statistical modeling. We found that using the intersections of the lines connecting surface features to determine the locations of the ground control points achieved good performance. It avoided the interferences from elevation changes at the surface features when extracting the elevation of GCPs from Shuttle Radar Topography Mission global 1 arc second data. It was a simple and effective way to obtain canopy height samples by selecting canopy samples and ground samples through manual interpretation and calculating their elevation differences. The red band and green band were selected by a stepwise method in the establishing of a multiple linear regression model. The validation of the model-predicted canopy height indicated a coefficient of determination of 0.72 and a root mean square error of 1.58 m. The proposed method extends the application of ZY3-02 data to mapping canopy heights of pure plantations in plain areas.

Author Contributions: Conceptualization, C.C.; data curation, M.L., W.C., and X.W.; funding acquisition, C.C.; investigation, M.L.; methodology, M.L.; project administration, W.C. and X.W.; software, X.W.; supervision, C.C.; writing—original draft, M.L.; writing—review & editing, C.C. and W.C.

Funding: This research was funded by the National Key Research and Development Program of China, grant numbers 2017YFD0600903 and 2016YFB0501505.

Acknowledgments: ZY3-02 data was provided courtesy of the China Centre for Resources Satellite Data and Application (CRESDA). The SRTMGL1 Version 3 product was retrieved from the online Data Pool courtesy of the NASA Land Processes Distributed Active Archive Center (LP DAAC) and the USGS/Earth Resources Observation and Science (EROS) Center.

Conflicts of Interest: The authors declare no conflict of interest.

References

1. Drake, J.B.; Dubayah, R.O.; Knox, R.G.; Clark, D.B.; Blair, J.B. Sensitivity of large-footprint lidar to canopy structure and biomass in a neotropical rainforest. *Remote Sens. Environ.* **2002**, *81*, 378–392. [CrossRef]
2. Garcia, O. Dynamical implications of the variability representation in site-index modelling. *Eur. J. For. Res.* **2011**, *130*, 671–675. [CrossRef]
3. Lim, K.; Treitz, P.; Wulder, M.; St-Onge, B.; Flood, M. LiDAR remote sensing of forest structure. *Prog. Phys. Geogr.* **2003**, *27*, 88–106. [CrossRef]
4. Bolton, D.K.; Coops, N.C.; Wulder, M.A. Investigating the agreement between global canopy height maps and airborne LiDAR derived height estimates over Canada. *Can. J. Remote Sens.* **2013**, *39*, S139–S151. [CrossRef]
5. Lefsky, M.A.; Keller, M.; Pang, Y.; De Camargo, P.B.; Hunter, M.O. Revised method for forest canopy height estimation from Geoscience Laser Altimeter System waveforms. *J. Appl. Remote Sens.* **2007**, *1*, 1–18.
6. Lefsky, M.A. A global forest canopy height map from the Moderate Resolution Imaging Spectroradiometer and the Geoscience Laser Altimeter System. *Geophys. Res. Lett.* **2010**, *37*, 1–5. [CrossRef]
7. Simard, M.; Pinto, N.; Fisher, J.B.; Baccini, A. Mapping forest canopy height globally with spaceborne lidar. *J. Geophys. Res. Biogeosci.* **2011**, *116*, G04021. [CrossRef]
8. Huang, H.; Liu, C.; Wang, X.; Biging, G.S.; Chen, Y.; Yang, J.; Gong, P. Mapping vegetation heights in China using slope correction ICESat data, SRTM, MODIS-derived and climate data. *ISPRS J. Photogramm. Remote Sens.* **2017**, *129*, 189–199. [CrossRef]
9. Hansen, M.C.; Potapov, P.V.; Goetz, S.J.; Turubanova, S.; Tyukavina, A.; Krylov, A.; Kommareddy, A.; Egorov, A. Mapping tree height distributions in Sub-Saharan Africa using Landsat 7 and 8 data. *Remote Sens. Environ.* **2016**, *185*, 221–232. [CrossRef]
10. Neuenschwander, A.; Pitts, K. The ATL08 land and vegetation product for the ICESat-2 Mission. *Remote Sens. Environ.* **2019**, *221*, 247–259. [CrossRef]
11. Qi, W.; Dubayah, R.O. Combining Tandem-X InSAR and simulated GEDI lidar observations for forest structure mapping. *Remote Sens. Environ.* **2016**, *187*, 253–266. [CrossRef]
12. Karjalainen, M.; Kankare, V.; Vastaranta, M.; Holopainen, M.; Hyyppä, J. Prediction of plot-level forest variables using TerraSAR-X stereo SAR data. *Remote Sens. Environ.* **2012**, *117*, 338–347. [CrossRef]

13. Capaldo, P.; Nascetti, A.; Porfiri, M.; Pieralice, F.; Fratarcangeli, F.; Crespi, M.; Toutin, T. Evaluation and comparison of different radargrammetric approaches for Digital Surface Models generation from COSMO-SkyMed, TerraSAR-X, RADARSAT-2 imagery: Analysis of Beauport (Canada) test site. *ISPRS J. Photogramm. Remote Sens.* **2015**, *100*, 60–70. [CrossRef]

14. Balzter, H.; Rowland, C.S.; Saich, P. Forest canopy height and carbon estimation at Monks Wood National Nature Reserve, UK, using dual-wavelength SAR interferometry. *Remote Sens. Environ.* **2007**, *108*, 224–239. [CrossRef]

15. Solberg, S.; Astrup, R.; Breidenbach, J.; Nilsen, B.; Weydahl, D. Monitoring spruce volume and biomass with InSAR data from TanDEM-X. *Remote Sens. Environ.* **2013**, *139*, 60–67. [CrossRef]

16. Zhang, Y.; He, C.; Xu, X.; Chen, D. Forest vertical parameter estimation using PolInSAR imagery based on radiometric correction. *ISPRS Int. J. Geo-Inf.* **2016**, *5*, 186. [CrossRef]

17. Ghasemi, N.; Tolpekin, V.; Stein, A. A modified model for estimating tree height from PolInSAR with compensation for temporal decorrelation. *Int. J. Appl. Earth Obs.* **2018**, *73*, 313–322. [CrossRef]

18. Cloude, S.R.; Papathanassiou, K.P. Polarimetric SAR interferometry. *IEEE Trans. Geosci. Remote Sensing* **1998**, *36*, 1551–1565. [CrossRef]

19. Yamada, H.; Yamaguchi, Y.; Rodriguez, E.; Kim, Y.; Boerner, W.M. Polarimetric SAR interferometry for forest canopy analysis by using the super-resolution method. In Proceedings of the 2001 IEEE International Geoscience and Remote Sensing Symposium (IGARSS), Sydney, Australia, 9–13 July 2001; pp. 1101–1103.

20. Papathanassiou, K.P.; Hajnsek, L.; Moreira, A.; Cloude, S.R. Interferometric SAR polarimetry using a passive polarimetric microsatellite concept. In Proceedings of the 2002 IEEE International Geoscience and Remote Sensing Symposium (IGARSS), Toronto, Canada, 24–28 June 2002; pp. 826–828.

21. Cloude, S.R.; Papathanassiou, K.P. Three-stage inversion process for polarimetric SAR interferometry. *IEE Proc. Radar Sonar Navig.* **2003**, *150*, 125–134. [CrossRef]

22. Le Toan, T.; Quegan, S.; Davidson, M.W.J.; Balzter, H.; Paillou, P.; Papathanassiou, K.; Plummer, S.; Rocca, F.; Saatchi, S.; Shugart, H.; et al. The BIOMASS mission: Mapping global forest biomass to better understand the terrestrial carbon cycle. *Remote Sens. Environ.* **2011**, *115*, 2850–2860. [CrossRef]

23. Moreira, A.; Krieger, G.; Hajnsek, I.; Papathanassiou, K.; Younis, M.; Lopez-Dekker, P.; Huber, S.; Villano, M.; Pardini, M.; Eineder, M.; et al. Tandem-L: A Highly Innovative Bistatic SAR Mission for Global Observation of Dynamic Processes on the Earth's Surface. *IEEE Geosc. Rem. Sens. M.* **2015**, *3*, 8–23. [CrossRef]

24. Leberl, F.; Irschara, A.; Pock, T.; Meixner, P.; Gruber, M.; Scholz, S.; Wiechert, A. Point clouds: Lidar versus 3D vision. *Photogramm. Eng. Remote Sens.* **2010**, *76*, 1123–1134. [CrossRef]

25. White, J.C.; Wulder, M.A.; Vastaranta, M.; Coops, N.C.; Pitt, D.; Woods, M. The utility of image-based point clouds for forest inventory: A comparison with airborne laser scanning. *Forests* **2013**, *4*, 518–536. [CrossRef]

26. Aguilar, M.A.; Del Mar Saldaña, M.; Aguilar, F.J. Generation and quality assessment of stereo-extracted DSM from GeoEye-1 and WorldView-2 imagery. *IEEE Trans. Geosci. Remote Sens.* **2014**, *52*, 1259–1271. [CrossRef]

27. Herrero, H.M.; Felipe, G.B.; Belmar, L.S.; Hernandez, L.D.; Rodriguez, G.P.; Gonzalez, A.D. Dense Canopy Height Model from a low-cost photogrammetric platform and LiDAR data. *Trees-Struct. Funct.* **2016**, *30*, 1287–1301. [CrossRef]

28. Immitzer, M.; Stepper, C.; Böck, S.; Straub, C.; Atzberger, C. Use of WorldView-2 stereo imagery and National Forest Inventory data for wall-to-wall mapping of growing stock. *For. Ecol. Manage.* **2016**, *359*, 232–246. [CrossRef]

29. Véga, C.; St-Onge, B. Height growth reconstruction of a boreal forest canopy over a period of 58 years using a combination of photogrammetric and lidar models. *Remote Sens. Environ.* **2008**, *112*, 1784–1794. [CrossRef]

30. Cohen, W.B.; Spies, T.A. Remote sensing of canopy structure in the pacific northwest. *Northwest Environ. J.* **1990**, *6*, 415–418.

31. Cohen, W.B.; Spies, T.A. Estimating structural attributes of Douglas-fir/western hemlock forest stands from Landsat and SPOT imagery. *Remote Sens. Environ.* **1992**, *41*, 1–17. [CrossRef]

32. Hudak, A.T.; Lefsky, M.A.; Cohen, W.B.; Berterretche, M. Integration of lidar and Landsat ETM+ data for estimating and mapping forest canopy height. *Remote Sens. Environ.* **2002**, *82*, 397–416. [CrossRef]

33. Xu, K.; Jiang, Y.; Zhang, G.; Zhang, Q.; Wang, X. Geometric Potential Assessment for ZY3-02 Triple Linear Array Imagery. *Remote Sens.* **2017**, *9*, 658. [CrossRef]

34. Farr, T.G.; Rosen, P.A.; Caro, E.; Crippen, R.; Duren, R.; Hensley, S.; Kobrick, M.; Paller, M.; Rodriguez, E.; Roth, L.; et al. The Shuttle Radar Topography Mission. *Rev. Geophys.* **2007**, *45*, 1–43. [CrossRef]

35. Hu, Z.; Peng, J.; Hou, Y.; Shan, J. Evaluation of recently released open global digital elevation models of Hubei, China. *Remote Sens.* **2017**, *9*, 262. [CrossRef]
36. Hirschmuller, H. Stereo Processing by Semiglobal Matching and Mutual Information. *IEEE Trans. Pattern Anal. Mach. Intell.* **2008**, *30*, 328–341. [CrossRef] [PubMed]
37. Richter, R.; Schläpfer, D.; Müller, A. An automatic atmospheric correction algorithm for visible/NIR imagery. *Int. J. Remote Sens.* **2006**, *27*, 2077–2085. [CrossRef]
38. Williams, M.S.; Bechtold, W.A.; Labau, V.J. Five instruments for measuring tree heights: An evaluation. *South. J. Appl. For.* **1994**, *18*, 76–82.

MDPI
St. Alban-Anlage 66
4052 Basel
Switzerland
Tel. +41 61 683 77 34
Fax +41 61 302 89 18
www.mdpi.com

ISPRS International Journal of Geo-Information Editorial Office
E-mail: ijgi@mdpi.com
www.mdpi.com/journal/ijgi